DATE

Invisible Matter
and the
Fate of the Universe

OTHER RECOMMENDED BOOKS BY BARRY PARKER

CREATION
The Story of the Origin and Evolution of the Universe

EINSTEIN'S DREAM
The Search for a Unified Theory of the Universe

SEARCH FOR A SUPERTHEORY
From Atoms to Superstrings

Invisible Matter and the Fate of the Universe

Barry Parker

Drawings by
Lori Scoffield

Plenum Press • New York and London

Library of Congress Cataloging in Publication Data

Parker, Barry R.
   Invisible matter and the fate of the universe / Barry Parker; drawings by Lori
Scoffield.
      p.   cm.
   Includes bibliographical references and index.
   ISBN 0-306-43294-3
   1. Cosmology—Popular works. 2. Dark matter (Astronomy)—Popular works. I.
Title.
QB982.P37   1989   89-8775
523.1—dc20   CIP

© 1989 Barry Parker
Plenum Press is a Division of
Plenum Publishing Corporation
233 Spring Street, New York, N.Y. 10013

Printed in the United States of America

Preface

Contemplation of the fate of the universe, and of course the end of the world, reminds me of the story of a traveler in the Swiss Alps. After hiking high into the mountains he and his guide spent the night in a chalet. Early the next morning he was awakened by a resounding crash, followed by a deafening roar and several loud cracks. Jumping out of bed in fright he ran to his guide and yelled, "What is happening? Is the world coming to an end?" The guide calmly sat up in his bed. "No . . . don't worry . . . it's just the rays of the sun. When they touch the snow on the peak of the mountain they cause it to hurl down into the valley. Then the rays warm the glacier, causing it to crack. It's not the end of the world, it's the dawn of a new day."

In the latter part of the book I will also be talking about the end of the world, and the fate of the universe. But, like the Swiss guide, I'd like to assure you that there is nothing to worry about: barring something unforeseen, no serious astronomical catastrophe will take place for millions of years. Still, the final fate of the universe is something we would like to know about, and understand. What will eventually happen to it? The key, it seems, is the amount of matter it contains. And, as you will see, part of this matter is invisible, or is at least hidden from us in some strange form. Scientists refer to it as dark matter.

In an earlier book (*Creation*) I described the birth and evolution of the universe up to the present. I like to think of the present book as the continuation of this story. The fate of the universe, however, depends on the amount of dark matter in it, and as a result, much of the book is directed to this topic. As you read it you will find that scientists are still unable to identify this dark matter. Many challenging problems remain. But, of course, that is what science is all about.

In writing a popular science book about a technical subject it is difficult to avoid all scientific terms. I have had to use a few, and have defined them as they arose. For those unfamiliar with such terms I have included a glossary at the end of the book. Another difficulty for some may be the notation I use for very large and very small numbers. To write them out in detail would be cumbersome, so I have used scientific notation. In this notation a number such as 10,000 is written 10^4 (the index is the number of zeros after the one). You may also be unfamiliar with the temperature scale I use, namely the Kelvin scale (K). On this scale the lowest temperature in the universe is 0 K (this is $-459°$ F on the Fahrenheit scale).

I am particularly grateful to the scientists who assisted me. Interviews were conducted, in most cases by telephone or letter, with many of the people mentioned in the book. I would like to thank them both for the interviews, and in some cases for photographs and reprints. They are: David Schramm, Jim Peebles, Jeremiah Ostriker, Halton Arp, Jan Oort, Gerhard Börner, Julius Wess, Vera Rubin, Craig Hogan, James Applegate, Virginia Trimble, Michael Turner, Geoffrey Burbidge, Richard Price, Marc Davis, Simon White, Edwin Turner, Anthony Tyson, John Bahcall, Joseph Silk, Freeman Dyson, Don Page, Randall McKee, Scott Tremaine, C. S. Frenk.

The sketches, cartoons, paintings, and some of the line drawings were done by Lori Scoffield. The remaining line drawings were done by Shawn Jensen. I thank them both for an excellent job. I would also like to thank Linda Greenspan Regan, Victoria Cherney, and the staff of Plenum for their assistance in bringing the book to its final form. And finally I would like to thank my wife for her support while the book was being written.

Barry Parker

Contents

CHAPTER 1

Introduction

As a teenager I built a reflecting telescope. After many hours of grinding a mirror, figuring, and assembling, I was ready for my first look at the heavens. I pointed it toward the Milky Way and began scanning the star clouds. It was a moment I'll never forget. A sense of elation overcame me; the stars were like tiny diamonds—some solitary, others in pairs and threes. Some were blue, others faintly red. I remember thinking as I looked at them that any one of them could have a planet orbiting it, perhaps with a civilization on it. I tried to imagine what they would be like.

I continued scanning the stars for a while; then, using a star map, I located a galaxy. It was a delightful sight—here was a system like our own Milky Way, consisting of billions of stars. It was a spiral, although I can't say for sure that I could see its spiral arms.

Over the next few evenings I found clusters of stars, nebulae (gaseous regions), and I observed each of the visible planets. Viewing them made me realize how insignificant Earth really is. As strange as it might seem, our sun is just an average-sized yellow star, with little to distinguish it from the billions of other stars in our galaxy. There are thousands just like it. Furthermore, there are stars that are thousands of times larger, and a few that are hardly bigger than Earth. And there are many that are much more colorful. Nevertheless, our sun is the ideal star for us. It supplies us directly or indirectly with everything we need for survival. If it were very much different we wouldn't be here.

A "cloud" of stars in the Milky Way.

A gaseous nebula.

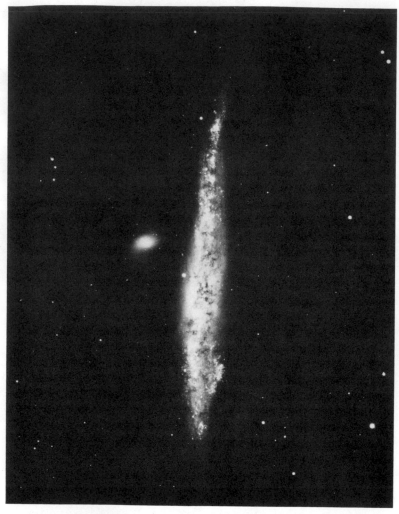

A galaxy (seen edge-on). (Courtesy National Optical Astronomy Observatories.)

An irregular galaxy. (Courtesy National Optical Astronomy Observatories.)

Even our galaxy (the Milky Way) is undistinguished; there are millions of galaxies similar to it in the universe. Yet it contains an amazing congregation. Besides star clusters, gaseous lagoons, and planets, there are pulsating stars—some that expand and contract over weeks and months. Some that appear to blink off and on in seconds. And there are giant stars that die in catastrophic explosions—supernovae. Most mysterious of all, though, are the bizarre black holes—tiny black spheres that trap everything that falls into them.

Closer to home we find asteroids—huge rocks flying through space at incredible speeds, and comets with long elegant tails of gas and dust. And around most of the planets are moons, some covered with ice, others with gigantic volcanoes. Each of these objects is bathed in a continuous wind of particles and radiation. Radio waves, infrared (IR) and ultraviolet rays, X rays and gamma rays, and particles of many different types make up this wind. We cannot see it with ordinary optical telescopes; specialized instruments such as radio telescopes, IR telescopes, and so on are needed.

The universe is, indeed, an amazing place: the incredible variety, the unbelievable distances involved are mind-boggling. Many of its mysteries may never be solved. One of the most bizarre of these mysteries came to light a few years ago: there is strong evidence that we are seeing only a small fraction of the universe. Over 90 percent of our galaxy, and the entire universe, seems to be invisible. We have evidence of its existence, yet we cannot see it, or detect it directly. Astronomers were shocked when they first realized this. It was almost as if someone announced we were seeing only 10 percent of Earth. The other part was there, but it was as yet undetected. You would no doubt scoff at such a suggestion. And at first astronomers were skeptical; most were certain that the "invisible matter" would be found as soon as better instrumentation was available. But the mystery remained, and it's still with us today.

But if we can't detect it directly, how do we know it's there? Most of the evidence for its existence comes from

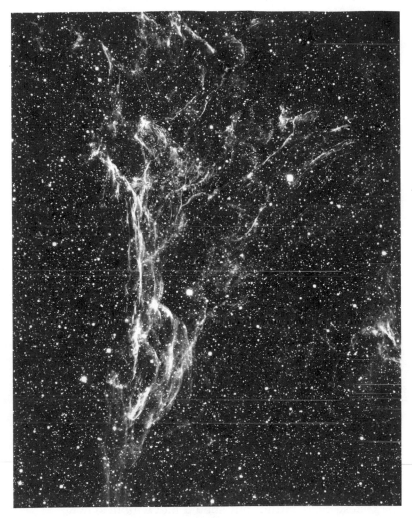

Evidence indicates there is "invisible" matter between the stars in this photo. (Courtesy National Optical Astronomy Observatories.)

inconsistencies in observations. The outer regions of galaxies, for example, rotate faster than they should. Galaxies in clusters move faster than they should. And the amount of matter in the universe itself is inconsistent with certain theories.

One of the first difficulties was noticed in relation to our galaxy. It is shaped like a disk, and although most stars orbit within this disk, a few stray above and below it. They do not stay out there for long, though, before they are pulled back by the powerful gravitational pull of the disk. A study of these stars shows that the gravitational pull is much greater than it should be. Putting it another way, we can say that for its strength there should be considerably more matter than we see. Part of our galaxy is therefore invisible—made up of "dark matter."

Also, there is a difficulty with the way stars move in the outer regions of the disk. This is best explained by comparing our galaxy to the solar system. They are, in many ways, quite similar: just as the planets of the solar system orbit the sun, the stars of our galaxy orbit its core. We understand the motions of the planets much better than we do the motions of the stars; a comparison is therefore useful in trying to predict what stars will do. Of particular importance, we know that the inner planets orbit faster than the outer ones; this was shown by Johannes Kepler many years ago, and is now referred to as Keplerian motion. We would therefore expect something similar in the case of the stars of our galaxy. Granted, things will be a little different: in the solar system almost all the mass is concentrated in the sun. In our galaxy it is distributed throughout it. Yet, if you look at a photograph of a galaxy it is brightest at the center, which seems to indicate the density of stars is greatest here, and therefore most of the mass is here. If this is the case, the gravitational attraction should be strongest near the core and the stars just outside this region should travel the fastest. Stars farther out should move more slowly.

But when observations of the outer stars and gas clouds were made, astronomers were shocked. They were traveling just as fast as stars closer to the core. This could only happen if there was a considerable amount of mass in the outer regions of our galaxy that we cannot see. Indeed, astronomers are now convinced that our galaxy is embedded in a huge cloud of dark matter that extends far beyond the most distant visible stars. But what form does the dark matter take? What is it? We don't know for sure; all we can really do is guess, and there have been plenty of guesses: brown dwarfs (small objects that didn't quite become stars), red dwarfs, white dwarfs, black holes, exotic particles, and many other things.

What about other galaxies? The problem does, indeed, exist for all galaxies. And the galaxy doesn't have to be a spiral. Elliptical galaxies also appear to have dark matter in them.

On a larger scale there is an even more serious problem. Our galaxy, the Milky Way, is part of a group of about 30 galaxies known as the Local Group. Observations and measurements of individual motions of members of the group show that it doesn't have enough matter to keep it gravitationally bound. The speeds of the individual galaxies within the cluster are so great that the entire group should be flying apart. Yet, it isn't.

And there are larger clusters such as the Coma cluster (in the constellation Coma), the Hercules cluster, and the Virgo cluster that contain thousands of galaxies. In their case, 50 to several hundred times more matter than astronomers now detect is needed to hold them together.

The dark matter mystery has become so serious that it now has top priority in cosmology. Thousands of astronomers are working on some aspect of it. A vast array of instruments—radio telescopes, X-ray telescopes, gamma-ray telescopes, computers—have been brought to bear on the problem. But why, we might ask, is there so much interest? The reason is simple: until it is solved, we will not be able to solve other problems such as how galaxies form, and how clusters and superclusters form. And perhaps the most important problem of all: we will not know the fate of our universe.

The first three-quarters of this book is devoted to dark

One of the major tools in the search for dark matter: radio telescopes. (Courtesy National Radio Astronomy Observatory.)

Domes at Kitt Peak Observatory. (Courtesy National Optical Astronomy Observatories.)

Earth millions of years in the future. The sun is only a dim star in the sky.

matter. We will consider the evidence for its existence and look at the various dark matter candidates. We will also look at how dark matter is important in relation to the structure in the universe, and how it relates to the gravitational lens. In the last one-quarter we will turn to the fate of the universe and show that it is determined for the most part by the dark matter.

What will the fate of the universe be? To answer this we

must look to the depths of space with our largest telescopes. Here we see galaxies, billions of them, all fleeing from us. The farther away they are, the faster they are traveling. The universe is expanding—and it appears to be expanding away from us. But looks are deceiving: it's not expanding away from just us (if it were we would be in an extremely important position; we would be at the center of the universe). If we went to another galaxy and observed the galaxies around it we would find that they were also moving away. Regardless of where you were in the universe the galaxies would be moving away from you. This is because it's the space between the galaxies that is expanding; the galaxies themselves aren't changing (except for their usual evolution). A useful analogy is an unbaked loaf of bread with raisins in it. When the bread is baked it expands and the raisins move away from one another, yet they stay the same size.

The galaxies are therefore expanding away from one another, but they themselves are not increasing in size. There is, however, a mutual gravitational attraction between them, and this causes them to slow down, or decelerate. If the pull is great enough they may actually stop. Since the gravitational pull depends on the number of galaxies, or more explicitly the amount of mass present, they will stop only if the average density of matter is above a certain critical density.

If the expansion stops, the universe is said to be closed. In this case the galaxies will collapse back on themselves and the universe will end in a "big crunch" of matter and radiation. The other possibility is that there is insufficient matter to stop the expansion. In this case the universe will expand forever, and is said to be open.

Which of these two cases applies to the universe? We still don't know for sure. In the last chapters of the book we will look at what will eventually happen to the universe in each of them. We will also look at what will eventually happen to Earth, and civilization in general. But first we must fill in a little background. In the next chapter I'll begin with the discovery of the expansion of the universe.

CHAPTER 2

The Expanding Universe

Our knowledge of the structure of the universe has come to us slowly, over many decades. Even as late as 1920 astronomers were still uncertain of what it looked like on the largest scale. At that time the structure of the solar system was fairly well established, and astronomers knew the distance to some of the stars. Furthermore, they knew that we lived in a large "system" of stars. But what did the universe look like beyond this system? How were the stars arranged?

One of the first to speculate on the overall structure of the universe was Thomas Wright. Born in 1711, Wright became interested in astronomy at an early age. And it was, perhaps, this interest that helped him overcome several handicaps. As a youth he had a serious speech impediment that restricted his formal schooling. But he refused to let it hold him back, and he began studying on his own. He soon mastered astronomy, mathematics, navigation, and surveying. But even here there was a problem: his father had little use for books and would burn them whenever he caught him with any. Wright therefore had to do much of his studying in secret.

Despite the handicaps, Wright managed to educate himself. And eventually he began to teach. He also began to publish pamphlets and books outlining his ideas. His best known book, *An Original Theory or New Hypothesis of the Universe*, was published in 1750. Although it was generally ignored at the time, it had a strong influence on later astronomers, and is important in that it was one of the first hypotheses for the structure of the universe that was approximately correct. Galileo had shown that the diffuse stream of light that shone overhead—the Milky Way—was composed of millions of individual stars. Wright became convinced that this was a "system" of stars—our system. Its appearance in the sky led him to postulate that it had the shape of a grinding wheel. Actually, it would have been a rather strange grinding wheel, as Wright was convinced that the stars went on forever. The diameter of the wheel would therefore have been infinite.

We know today that Wright's model was fairly close to reality. We do, indeed, live in a disk-shaped system of stars, the Milky Way. We see many stars when we go out on a summer evening and look up in the direction of the Milky Way because we are looking along the disk. If we look at 90 degrees to this we would expect to see few stars, and we do.

For years astronomers believed that we lived near the center of this system. After all, the Milky Way appears to surround us; we see it throughout the year. We now know, though, that this isn't the case. We actually live about three-fifths of the way out from the center.

Wright's ideas, although ingenious in some ways, were nevertheless vague and unclear. In fact, he not only formulated the grinding wheel model, but later in life put forward several other models that were quite different—some of them rather weird. And perhaps because of this his work was generally ignored in his day.

One person who didn't ignore it, however, was Immanuel Kant. Although best known as a philosopher, Kant made important contributions to astronomy. Born in 1724 to a poor family of Lutherans, Kant lived his entire life in his birthplace, Königsberg (in old Prussia). Indeed, during his entire life he never ventured farther than 60 miles from the city. His mother died when he was 13, and at 16 he entered the University of Königsberg to study theology. Never as devoutly religious as his parents, he soon found that he preferred the mathematics and physics classes to the theology ones. He obtained his doctorate in 1755, and in the same year published his most important scientific work, *General History of Nature and Theory of the Universe.* He had read about Wright's model, not from Wright's book, but from a newspaper article. And fortunately, whoever wrote the article used only the kernel of Wright's arguments, so the article was much clearer than the book. There is, in fact, no evidence that Kant ever saw Wright's book. Nevertheless, he soon became intrigued with his ideas. Astronomers had already begun to notice fuzzy white patches in the telescope, objects that soon became known as white nebulae. Wright had hinted that they might be systems similar to our own but lying at great distances beyond it. But again his ideas were hazy. Kant accepted the notion that our system, the Milky Way, was a disk-shaped object composed of millions of individual stars. But he went on in his book to argue that the white nebulae were similar systems of stars—"island universes of stars," he called them. The name stuck, and is still used to some extent today. Kant was sure that the white nebulae were composed of stars, but his ideas were not immediately accepted. Other scientists were equally sure that they were gaseous objects within the Milky Way.

What was needed now was a catalogue of these nebulous objects. Interestingly, the first catalogue that was published was not made so that astronomers could easily find and study these objects; it was made so that they could avoid them. The astronomer who developed it was born 6 years after Kant; his name was Charles Messier. When he was 12 his father died and he was raised by an older brother. His interest in astronomy was sparked at age 18 when he witnessed an eclipse of the sun. He was so awed by it that he decided then and there to become an astronomer, and at 20 he went to the Paris Observatory as an apprentice. Soon after arriving he learned that everyone was excited about the possible return of Halley's Comet in 1758. Seventy-four years earlier Halley had predicted its return; Messier knew that his reputation would be established if he could find it first. He searched diligently for 18 months, but unfortu-

Charles Messier.

nately was looking in the wrong region of the sky and a French peasant saw it before him. Messier was heartbroken over his failure, but soon began a lifelong love for comets.

A major problem for comet hunters at the time were the fuzzy objects in the sky that did not move. When a comet is first spotted in a telescope it usually looks like a small fuzzy ball. Unfortunately, the white nebulae and other objects also look like small fuzzy balls. A comet can be distinguished from them only if it moves relative to the background stars, and it usually takes several nights to determine if this is the case. To Messier the white nebulae (and other similar-looking objects) were therefore nuisances. He decided to make up a table of these nuisances so that others would not have to spend time watching to see if they moved. In 1784 he published his table; it contained 103 entries. We now refer to most of the well-known white nebulae by their number in this catalogue (e.g., M31, M32).

Messier's later life was one of considerable misery. He accidentally fell down a dark shaft and lay for hours with broken bones before anyone found him. He suffered pain from this accident for the rest of his life. Furthermore, despite his contributions, he was not taken seriously by his colleagues. He had considerable difficulty getting financial support, and in getting elected to the French Academy of Science. And although he did discover a number of comets, he is not remembered for any of them. But he is remembered for this table.

HERSCHEL AND COMPANY

Messier's telescopes were small, crude, and their resolving ability was poor. Larger telescopes were obviously needed if progress was to be made. And they soon came. The "father" of the large telescope, William Herschel, and his brother Jacob arrived in London from Germany in 1757. Although almost penniless, William, an accomplished musician, soon found work giving music lessons. And in time, as his reputation increased, he began playing for the public, directing, and composing. His ambition during this period of his life was to become a famous composer.

Two years after arriving in London, Jacob returned to Germany and William was left alone. Then in 1767 their father, who was still in Germany, died. Caroline, their younger sister, who had been living with her father, longed to visit William. She had few friends, and because smallpox had left her scarred she was shy and tended to shun people. William had always been her favorite brother and when he asked her to come to England she eagerly accepted.

She soon became his assistant, copying music and caring for him. In 1773 an important change in his life began. He bought a copy of Ferguson's *Astronomy*, a popular-level book describing the stars and planets. Finding it fascinating he read it again and again, taking it to bed almost every night, falling asleep with it lying on top of him. He wanted to explore the stars, but needed a telescope. Soon he was studying books on optics and experimenting with lenses and mirrors.

William Herschel.

The next few years was a period of gradual transition. More and more of his time was spent building telescopes and observing, and less and less time was spent with music. He was now about 40 years old, so it can't be said that he got an early start. Despite this, throughout the rest of his life he probably spent more time at the eyepiece of a telescope than anyone who has ever lived.

An important turning point occurred in 1781. During a routine scan of the sky he came across a fuzzy object that appeared to be a comet. He watched it night after night, but it didn't move like a comet. In time he determined its orbit. This convinced him that it was a planet—a previously undiscovered planet. Other astronomers were also soon convinced. Thus he became the first astronomer to discover a planet. Herschel suggested that it be named after George III, the King of England. But it was eventually named Uranus, after the father of Saturn. Herschel became famous literally overnight, and within a short time he was given a stipend from the King. His financial worries were now over; he could devote all his time to astronomy. Furthermore, he now had enough money to build even larger telescopes.

Most of Herschel's important astronomical work was done with a 20-foot (approximate length of the tube) reflecting telescope. (A reflecting telescope has a large mirror that reflects the light it gathers onto a small mirror where it can be observed with an eyepiece.) Although he later built a 40-foot monster, little of importance was accomplished with it. It was so large and awkward that it required several men to operate it. Furthermore, it was so dangerous that a number of people were hurt while working with it.

Herschel was fascinated with Messier's catalogue of nebulous objects, and soon became convinced that Messier had missed some—perhaps a dozen or so. And indeed Herschel showed that he had. Over a period of seven years he found 2500 more. Like Kant he was convinced they were island universes of stars outside the Milky Way. But of course he was unable to prove it; his telescopes, although the largest of the day, were incapable of resolving the nebulae into stars.

But Herschel was not only fascinated by the white nebulae, he was equally fascinated by our system, the Milky Way. How big was it? What was its shape? To answer these questions he undertook an extensive program of "star counting." He divided the sky into small squares and tediously counted the stars in each square, eventually coming to the conclusion that the Milky Way was pancake-shaped with the sun near the center. Although he drastically underestimated its size, this could be expected as astronomers still had little idea how large the universe was.

Herschel married rather late in life—much to the dismay of Caroline. He and his wife had one child—a son, John. John had little interest initially in following in his father's footsteps, but for many years was uncertain what he wanted to do. His major problem, perhaps, was that he was interested in so many things. William wanted him to go into the clergy, but he had little use for the church. Finally, although he was much more interested in mathematics at the time, he decided on law. Soon after graduation, though, he found that law bored him so he switched to teaching, but he also found teaching to be boring. At this point he still had no plans for astronomy. But a visit to his father on his deathbed changed this. He decided he would continue his father's work. William had mapped most of the northern hemisphere; John decided to continue the work in the southern hemisphere. So he packed his bags and the 20-foot telescope and took it to the Cape of Good Hope. And for the next four years he worked diligently extending his father's catalogue while studying comets, sunspots, and various other astronomical phenomena. Unfortunately, when he returned to England he left astronomy, never returning to it. For a while he worked on the new discovery of photography (in fact, he coined the name "photography"); then he wrote popular astronomy books and finally (alas!) he went into politics.

Herschel's successor, William Parsons, was born in 1800 to an aristocratic Irish family. He entered the English Parliament at 21 as a representative from Ireland, but soon realized his true interest lay in the building of complex devices and machinery. He also found that his position left him with considerable free time. After hearing about Herschel's astronomical discoveries he became determined to put this free time to good use. He would continue with Herschel's work; he would build even larger telescopes. But he had no idea how to build a telescope, as Herschel had said little about it in his writings. Parsons therefore began experimenting with large metal mirrors. Using alloys of tin and copper he eventually cast a 36-inch mirror.

His father died in 1841 and William became earl of Rosse. This gave him both the time and the money to construct a much larger telescope. Unlike most who would tackle something only slightly larger than 36 inches, Rosse decided on one twice as large. This was even larger than Herschel's monster. Undaunted, Rosse went ahead with the project, casting the mirror in 1842, then grinding and polishing it the following year. By 1844 the mirror was complete; it was then mounted between

William Parsons (earl of Rosse).

two concrete piers and moved up and down with chain cables and pulleys.

Despite the difficulties maneuvering the telescope, and the perpetually bad weather of Ireland, Rosse made good use of it. The white nebulae were still a major problem. What were they? Rosse turned his telescope to them and discovered that many of them had a strange spiral structure. He made drawings of them and presented them to the Royal Society.

Rosse was convinced that they were island universes of stars, but even his great telescope could not resolve them. So he kept his views to himself.

DAWN OF A NEW ERA

Although the white nebulae had been observed for many years little was known about them. It was clear that they had different structures: some had a spiral structure, others were elliptical, and a few were quite irregular. But what were they: island universes of stars, or just gaseous clouds in our galaxy? Was it, in fact, possible to find out?

Larger telescopes would obviously help but it was a series of discoveries that had taken place many years earlier that eventually led to the breakthrough. Newton made the initial discovery. He noticed that when light from the sun was passed through a prism it became a rainbow of colors—a spectrum. In 1802 William Woolaston, while examining this spectrum, noticed several dark lines in it. He thought they were boundary lines between the various colors of the spectrum and didn't study them in detail. This was left to the German physicist Joseph von Fraunhofer. Born in 1787, Fraunhofer was orphaned at 11 and nearly killed at 14 when the building he was in collapsed. Despite many hardships throughout his youth he taught himself optics and was soon making significant improvements in the optical instruments of the day. In 1814 he built a crude spectroscope and allowed sunlight to pass through it. Like Woolaston he saw dark lines, but where Woolaston only saw a few, he saw hundreds (over 600 in all). He carefully measured and catalogued their position. He then turned his spectroscope to the moon and planets, and saw similar lines in their spectrum. Indeed, he even went as far as observing the spectrum of a few stars. But their spectrum confused him; the lines were in a different position. He had come close to an important discovery, but didn't quite make it. And because scientists never took his work seriously it would be another 50 years before the importance of spectroscopy in astronomy was realized. Nevertheless, during this time another important discovery was made.

In 1842 the Austrian physicist Christian Doppler was experimenting with sound when he noticed that its pitch varied depending on whether the source of the sound was approaching or receding. As you likely know, sound is transferred as a vibration of the air molecules. When pitch varies it is due to a change in the vibrational rate, or frequency, of the sound. Doppler discovered that if the source was approaching, the frequency was higher than it was if the source was at rest; similarly, if it was receding it was lower. This is now known as the Doppler effect. You can easily observe it by standing near a highway and listening to the hum of cars as they pass. Better still, listen when someone blows a car horn at you. As the car passes there will be a distinct change (a lowering) in the horn's pitch.

Doppler worked out mathematical formulas for the change in frequency and in 1845 they were tested in a memorable experiment. Trumpeters stood on the open car of a passing train sounding their trumpets as musicians stood alongside the track listening to the sound. There was no doubt that the Doppler effect was valid.

Since light is also a wave, Doppler was convinced the same phenomenon would occur there. And two years later the French physicist Armand Fizeau showed that it did. The effect in the case of light, although much smaller than the corresponding effect for sound (at ordinary velocities), soon became just as easy to measure. The reason: the spectroscope.

Fraunhofer's work was taken up eventually by the German physicist Gustav Kirchhoff. Kirchhoff studied at Königsberg, graduating in 1847. Soon after graduation he teamed up with Bunsen, who had just developed what we now call the Bunsen burner. They began studying the spectrum of elements heated to incandescence, soon showing that each element gave its own characteristic spectrum (pattern of bright colored lines). Kirchhoff later noticed that some of the bright lines were in the same position as the dark lines Fraunhofer had studied.

Although Kirchhoff developed the techniques and much of the theory of spectroscopy, he did not apply them to astronomy. The first astronomical use of spectroscopy came from the wealthy English amateur William Huggins. Huggins built an observatory on the top of his house in London in 1859 and equipped it with an 8-inch telescope. He began studying sunspots, sketching Jupiter and Saturn, but soon yearned to do something more substantial. Then he heard of Kirchhoff's work with a spectroscope. Having been trained in chemistry he quickly realized its potential in astronomy. He therefore built a spectroscope, attached it to his telescope, and went to work. He noticed that while stars gave a spectrum of dark lines on a bright background, some of the white nebulae gave a bright line spectrum. One of his most important contributions was the discovery of the Doppler effect in the spectrum of stars. He noticed that the lines were shifted slightly in the case of several stars, and correctly interpreted the change as due to the star's motion relative to us. Huggins was also one of the first to experiment with photography in astronomy.

A new door had now been opened, and astronomers soon took advantage of it. With spectroscopy they could tell what elements were in the atmosphere of a star, what its surface temperature was, how fast it was moving away from us, and how fast it was spinning. Furthermore, within a few years the spectroscope would be used to study the white nebulae.

DISCOVERY OF THE EXPANSION

By 1900 it was clear that there were two kinds of white nebulae: those that exhibited dark line spectra, as stars did, and those that exhibited bright line spectra. But at this stage there were many problems, and astronomers were still uncertain what the white nebulae were. Furthermore, there were also problems in relation to our galaxy, the Milky Way. Astronomers were uncertain how large it was, and where we were located in it. (Most believed that we were at the center.) Then came another important discovery.

Henrietta Leavitt graduated from what is now Radcliffe College in 1892. In 1902 she joined the staff of Harvard Observatory, under William Pickering. Women were still considered incapable of doing serious research at that time, and most were given tedious, routine tasks. Large numbers of stellar spectra were obtained each night, and all had to be carefully classified. Pickering was also interested in variable stars (stars that vary in brightness). One of Leavitt's duties was to check the plates of an

Henrietta Leavitt.

irregular galaxy in the south called the Small Magellanic Cloud (SMC) for such variables. Plates were sent regularly to Harvard from a southern observatory in Peru.

The Small Magellanic Cloud is close enough that individual stars can be seen. Leavitt noticed that most of them varied periodically with a period of several days. She went on to show that those with the highest average brightness had the longest period. If they had been in our galaxy this wouldn't have been important because they would likely have all been at a different distance. But they were in the Small Magellanic Cloud, and therefore were all roughly at the same distance, much in the same way that the people of Los Angeles are roughly the same distance from the people of New York. This meant that when one star appeared brighter than another it actually was brighter. On the basis of this Leavitt suggested that there was a relationship between period and luminosity (brightness).

Leavitt published her results in 1912 and wanted to pursue the project, but Pickering thought she had already gone too far.

Ejnar Hertzsprung.

Such research was for men, she was told. So she returned to her routine projects.

Leavitt's work, however, soon came to the attention of the Danish astronomer Ejnar Hertzsprung. Hertzsprung noticed that the variables she was studying were similar to a type called Cepheids that had been identified a few years earlier by John Goodricke (they are named for the brightest of the group, delta Cephei, in the constellation Cepheus). He also realized that since it was likely that Cepheids were the same regardless of where they were, he would be able to determine their intrinsic, or "absolute," brightness as long as he could determine their period. All he would need is to determine the distance to a single one; this would, in effect, calibrate the scale. But none were close enough for a distance determination, so Hertzsprung resorted to a statistical method. Using it he determined the distance of the Small Magellanic Cloud to be 30,000 light-years (we now know this is about five times too small). Although it was still far from accurate, at this stage the Cepheid period–

luminosity relation would eventually play an important role in astronomy.

As it turned out, one of its first uses came, not in relation to the white nebulae, but in relation to our own galaxy. The astronomer who took advantage of the new technique was Harlow Shapley. Born on a farm in Missouri in 1885, Shapley had a spotty elementary education. When he was 15 he went to a business school, then became a crime reporter for the *Daily Sun* of Canute, Kansas, later moving to the *Joplin Times* in Joplin, Missouri. After a couple of years of reporting he began to realize he needed more education. He decided he wanted a college degree, but he hadn't even gone to high school, so he thought he'd better complete that first. With his younger brother John he applied at a nearby high school. "Sorry, you're not qualified," came the reply. Shocked but determined, they applied at another school, where they were accepted. Two semesters later Shapley graduated; in fact, he was valedictorian of his class—a class of three.

His aim now was to go on in journalism to the University of Missouri. But when he arrived at the university he discovered to his dismay that the journalism school wouldn't open until the following year. Determined not to lose a year he started thumbing through the course catalogue. On the first page was ARCHAEOLOGY. He couldn't pronounce it, although he thought he knew what it was. Paging on he came to ASTRONOMY. "I could pronounce that," he said. So he signed up for it, and in four years had both a bachelor's degree and a master's degree in it.

In the last year of his master's degree he got a fellowship to Princeton University. He was nervous when he arrived at Princeton; he would be working under the famous astronomer Henry Norris Russell. And for the first few weeks Russell made it difficult for him. "He intimated that I was a wild Missourian . . . and no one should expect much," Shapley said later. But they soon became good friends.

As he approached graduation he wrote to George Ellery Hale at Mt. Wilson in California about a position. Hale agreed to

Harlow Shapley.

meet with him in New York. Shapley went to New York the day
before his appointment, and having nothing to do that evening
he went to the opera. The next morning he had breakfast with
Hale, and during their conversation he casually mentioned that
he had attended the opera; Hale seemed interested and began
asking him questions about it. They continued talking about opera until the meal was over, then Hale excused himself and left.
Not a word was said about astronomy, or the job. Shapley was
in a state of shock. "What did I do wrong?" he asked one of his
professors when he got back to Princeton. Apparently he had
done nothing wrong, for within days he got a letter asking him
to come to Mt. Wilson. Hale had just been playing a little game
with him. Shapley's first project at Mt. Wilson involved globular
clusters—roughly spherical systems of from a few hundred
thousand to a few million stars. He discovered that most of
them contained Cepheids, and therefore with the use of the
period–luminosity relation he could determine their distance.
When he plotted their positions he found that they were
grouped in a sphere, but the sphere was not centered on Earth.

What did this mean? Shapley assumed that they were likely grouped symmetrically about the center of our galaxy, and if so it meant that our sun was off to one side of it. And, indeed, we now know that we are about three-fifths of the way to the edge. Shapley's calculated size for our galaxy, however, was about 2½ times too large. Furthermore, he was convinced that the white nebulae were much smaller than the Milky Way, and only satellite systems of it. Curtis, an observer at Lick Observatory, violently opposed Shapley's view. And in 1920 the two found themselves in debate at the AAS meeting in Chicago. This is now frequently referred to as the "Great Debate." Shapley, however, didn't think of it as a debate. "I don't think the word 'debate' was ever used," he said. "It was more of a symposium—a brief talk with a rebuttal." He said he was quite surprised when, years later, people started writing about it as if it were an important turning point in astronomy. He didn't remember it as such. To him it was just a "pleasant meeting."

After Shapley presented his views Curtis gave reasons for his belief that the white nebulae were systems of stars similar to the Milky Way. He was convinced, however, that we were at the center of our system. When the fog finally cleared—several years later—both men were right in some of their arguments and wrong in others.

It was the Mt. Wilson astronomer Edwin Hubble who in the end settled the debate. Using long exposures he photographed some of the nearby white nebulae and, after resolving their outer regions into stars, showed that some of them were Cepheids. This allowed him to determine their distance. And indeed it was soon clear that they were galaxies—island universes of stars—independent of ours. Some of them, in fact, were considerably larger than the Milky Way; they were definitely not satellite systems.

The first spectra of these systems had actually been taken years earlier. Vesto Slipher of Lowell Observatory in Flagstaff, Arizona, directed a 24-inch telescope equipped with a spectroscope toward the great white nebula in the constellation Andromeda in 1912 and found that it gave dark lines just as stars do. In fact, the lines were slightly shifted, indicating that it was approaching us. Slipher went on to obtain the spectra of other galaxies and found most of them were moving away from us. But it was left to Hubble to determine the meaning of Slipher's data. Using it, and other spectra obtained by his assistant, Milton Humason, he announced in 1929 that there was a relation between redshift and distance. The farther away the galaxy, the faster it was receding. The universe was expanding.

Oddly enough, the prediction had already been made several years earlier as a result of theoretical studies. Shortly after Einstein published his general theory of relativity in 1916 he went on to apply it to the universe. The cosmology he formulated was a static one—individual objects had velocities, but on the average the universe was stationary. Strangely, Einstein missed a solution that was quickly found by the Dutch astronomer Willem de Sitter. But de Sitter's universe was empty; furthermore, it was soon discovered that it was not stationary. If two objects were placed in it they would move away from one another. On the basis of this the prediction was made that the universe was expanding.

Both of these models were eventually shown to be incorrect. The model we accept today was published in 1921 by the Russian Aleksandr Friedmann. He showed that there were three options for the evolution of the universe, depending on how much matter it contained (i.e., its average density). If the average density was under a certain critical amount the universe would expand forever. If the average density was greater than the critical amount, on the other hand, the universe would eventually stop expanding and collapse back on itself. We will talk about this in more detail later.

CONTROVERSY

But are we certain that the redshift of spectral lines indicates an expansion of the universe? How can we find out for sure?

Determination of the age of the universe (the time since the big bang explosion) would obviously be helpful. Unfortunately, there has, in the last few years, been some controversy as to the age. For years the accepted age of the universe was 18–20 billion years. This was a value obtained painstakingly over many years by Allan Sandage of Hale Observatory, and Gustav Tammann of Switzerland. While Sandage and Tammann were working on their estimate, however, an astronomer at the University of Texas, Gerard de Vaucouleurs, was independently working on the same problem. And strangely, he was consistently getting a value about half theirs. At first, little attention was paid to de Vaucouleurs's value. Then in 1979 John Huchra of Harvard University, Marc Aaronson of the University of Arizona, and Jeremy Mould of Kitt Peak National Observatory announced that, using a particularly reliable method, they had obtained a value intermediate between that of Sandage and de Vaucouleurs. The announcement caused considerable confusion.

Was our estimate of the age of the universe really that inaccurate?

Let's begin with some background. As I mentioned earlier, Hubble made a plot of redshift (velocity) versus distance for galaxies. This told him how fast the universe was expanding, and it also gave him the approximate age of the universe. To see why, all you have to do is visualize the expansion as occurring in a movie, then ask yourself how long it would take if you reversed the film for the galaxies to get back together to the same point.

When Hubble first did this he got the embarrassingly low age of 2 billion years. This was embarrassing because geologists had already found rocks on Earth that they were sure were older than this. To see what the problem was, let's look at the plot Hubble made (distance versus redshift). Getting the redshift of the galaxy is straightforward enough. Extremely long exposures are needed if the object is dim but, in general, there are no serious inaccuracies. Determination of the distance to the galaxy, however, is a different matter. It is difficult, to say the least.

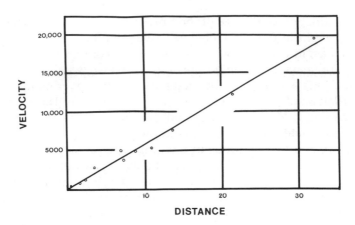

An early Hubble plot of velocity (redshift) versus distance.

Hubble determined the distance to the nearest galaxies using the Cepheid period–luminosity relation, but in more distant galaxies Cepheids were not visible so he switched to extremely bright supergiant stars (to a first approximation they were all about the same brightness). Finally, when no stars were visible he had to use the brightness of the galaxy itself. Again, to a rough approximation, galaxies of the same type are all of the same brightness and Hubble took advantage of that. Using a technique of this type—what he referred to as a "cosmic ladder"—he was able to step his way to the outer regions of the universe.

The first correction to Hubble's estimate came when astronomers discovered that the light from galaxies was being dimmed by intergalactic gas. When the correction was made the age of the universe doubled. Then it was discovered that there was more than one type of Cepheid variable and further corrections had to be made. Eventually the age was squeezed up to 10 billion years—easily consistent with geological results.

Sandage and Tammann, however, continued Hubble's work and over a period of 20 years, using better instruments and techniques, they determined the age to be 18–20 billion years. But as I mentioned earlier, de Vaucouleurs had also been working on the problem, using a cosmic ladder just as they had—in fact his cosmic ladder was somewhat more sophisticated. It included novae (exploding stars), Cepheids, supergiant stars, and other types of variable stars. And he consistently got an age of about 10 billion years.

Why the difference? The major reason, according to de Vaucouleurs, was because Sandage and Tammann were ignoring the intergalactic gravitational forces that act between galaxies. Astronomers discovered several years ago that our galaxy is part of a cluster of about 30 galaxies, a group we now call the Local Group. These galaxies are gravitationally bound and move through space together. Most galaxies, in fact, reside in groups such as this. de Vaucouleurs, however, went a step further: he showed that our Local Group is part of a cluster of clusters—a supercluster. He referred to this as our Local Supercluster. According to de Vaucouleurs the supercluster is dominated by a huge group near the center, called the Virgo cluster. And because of the mutual gravitational pull of the supercluster we are "falling" toward its center at a rate of about a million miles per hour (450 km/sec). This has a serious effect on how we see the expansion of the universe; it artificially lowers the rate, and as a result we see galaxies in the direction of Virgo appearing to recede slower from us than they should be. This, in turn, affects the calculated age of the universe. According to de Vaucouleurs, Sandage and Tammann do not take this into consideration.

Huchra, Aaronson, and Mould hoped to resolve the controversy. They used a technique that was a refinement of an earlier one used by the astronomers Brent Tully of the University of Hawaii and Richard Fisher of the National Radio Observatory. In this method the brightness of a galaxy is compared to its rotational speed. In effect, the rotational speed tells us how massive a galaxy is, which in turn gives us its absolute magnitude. We can then determine how far away the galaxy is by observing its brightness as seen from Earth.

When Tully and Fisher developed the method, they used visual brightness, which, as it turned out, was plagued with difficulties due to the obscuring medium within our galaxy. Huchra and his team switched to the infrared where the problems were minimal. Their first value, which was obtained using galaxies in a direction away from the Virgo supercluster, was intermediate between Sandage's and de Vaucouleurs's estimates. de Vaucouleurs then suggested they take measurements in the direction of Virgo, and when they did they got a number quite close to his. Still, Sandage and Tammann were unconvinced. They were certain that their measurements were more accurate and their age more representative of the actual age of the universe.

Is there any way of determining which of the two numbers is most representative? Indeed there is, and it is useful not only in helping resolve the difficulty but also in helping convince us that the redshift interpretation is correct. If several different methods of determining the age, for example, give roughly the same value, it is reasonable to suppose that the redshift interpretation (i.e., the universe is expanding) is valid. There are, it turns out, two other methods of determining the age of the universe. Actually, they both give the age of our galaxy, but since we believe our galaxy is just slightly younger than the universe, they also give a good estimate for the age of the universe. The first comes from a plot of stars known as the HR diagram. If we plot the absolute brightness versus the surface temperature for a large group of stars we get a broad column of points. In general they lie in a diagonal across the graph. By measuring the length of this diagonal we can determine the age of the group. If we make such a plot for globular clusters, for example, we get ages that range from 8 to 18 billion years. And since globular clusters are believed to have formed about the same time as our galaxy this gives the approximate age of our galaxy—and of the universe.

The second method is to observe the decay rate of radioactive materials. The half-life of all radioactive substances is

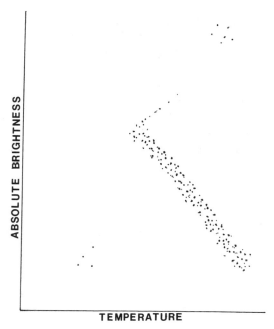

HR diagram of cluster showing breakoff point.

known, and by measuring the abundance of the radioactive atoms in the solar system we can determine its age. We get 7–18 billion years.

But again we have quite a range. What is the best value? David Schramm of the University of Chicago and several colleagues have attempted to answer this. They have taken all of these methods and used them to "squeeze in" on the best possible estimate for the age, and they came up with 15–16 billion years. Still, everyone is not satisfied, and we can therefore only say with certainty that our universe is somewhere between 10 and 20 billion years old. Furthermore, there are other controversies.

Certainly most astronomers believe that the redshift interpretation according to the Doppler effect is valid and that our universe is indeed expanding. There are a few, however, who doubt this interpretation. Halton Arp, known to his friends as Chip, is the leader of this group. Arp received his bachelor's degree from Harvard in 1949 and his Ph.D. from Caltech in 1953. After a brief stay at Indiana he went to Mt. Wilson and Palomar Observatories. Soon after he arrived he began working on peculiar galaxies and in 1966 published his *Atlas of Peculiar Galaxies*. As the name suggests, these are galaxies that are peculiar in some way—interacting with other galaxies, or distorted. While examining the photographs in his atlas he began to notice that certain types of radio sources, called quasars, seemed to be frequently associated with ordinary, but disturbed galaxies. This eventually led him to put forward the idea that the quasars had been ejected from the galaxy. His colleagues were astounded—calling his idea nonsense and refusing to take it seriously.

Then he heard of two objects—a quasar and a galaxy—that appeared to be very close together in the sky (called NGC 4319–Markarian 205). He took a four-hour exposure of them with the 200-inch telescope. "When I developed the photograph I was surprised and excited to find a luminous connection between the quasar and the galaxy," he said. He was particularly excited because earlier Daniel Weedman of Vanderbilt University had shown that the two objects had different redshifts. This meant that they couldn't be connected. If their recessional velocity was due to the big bang they had to be at different distances from us. Yet they appeared to be connected.

Other astronomers took photographs and stated they were certain the two objects were just superimposed on one another. Arp continued to uncover other cases of quasars apparently connected to galaxies. (He now has six.) But as he pushed his ideas, he had increasing difficulty publishing them. Astronomers did not take him seriously.

Arp is still convinced that there is no direct evidence for Doppler shifts in any extragalactic objects. I asked him, if the redshift is not due to the Doppler effect, what was causing it? "The only explanation I can see is that certain kinds of objects

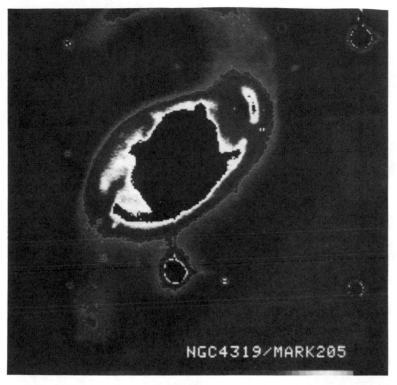

NGC 4319–Markarian 205. (Courtesy H. Arp, Max Planck Institute.)

(suggested to be younger) have intrinsic redshifts. Other redshift mechanisms such as redshifting photons by gravitational fields or interactions with a medium or spontaneous photon decay are possible, but not likely in my view," he said.

Arp has accumulated a large amount of evidence that he recently summarized in a book titled *Quasars, Redshifts and Controversies*. Most astronomers still do not take it seriously. Nevertheless, Arp is convinced he has enough "discordant" objects to cast a shadow on the Doppler interpretation. Is he right? I suppose we'll just have to wait and see.

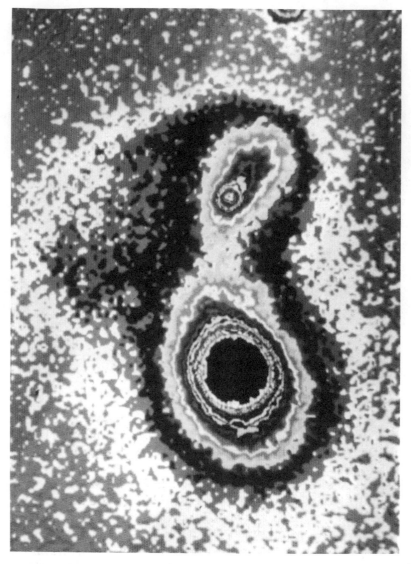

A quasar and a galaxy interacting. Computer enhanced.

Also unexplained is an effect discovered in 1976 by Will
Tifft of Steward Observatory. Tifft noticed that redshifts seemed
to be "clumped" around certain values. The effect is referred to
as the "quantization of redshifts," named for a similar effect that
occurs in the microworld. This happens not only for single
galaxies, but also for double galaxies and even multiple systems.
There is no known reason why they should do this. Certainly, if
galaxies are all moving away from us at random according to the
Doppler effect, you wouldn't expect this. Why does it happen?
Again, we don't know.

41

CHAPTER 3

The Missing Mass Mystery Unfolds

One of the major problems in astronomy is that of determining the weight, or mass, of objects in the universe. Astronomers can weigh stars relatively easily, but galaxies and clusters of galaxies are another matter. Let's look at how they weigh stars. The first requirement is a binary system—two stars that revolve one around the other. Fortunately, it's not difficult to find such systems, as many of the visible stars in the sky are binaries. Polaris, our North Star, for example, has a dim companion, and Sirius, the brightest star in the sky, also has a faint companion. In many cases we can't see the dimmer of the two stars, but we know it's there when we look at the spectral lines of the system: because of the Doppler effect they move back and forth as the two stars revolve around one another. From the period of revolution of the stars we can obtain a relation between the mass of the stars and their brightness (luminosity).

If we do this for a large number of binary systems we can make a plot of mass versus luminosity, then use the plot to determine the mass of any star of known luminosity. The technique is accurate, of course, only if the light emitted per unit mass for all the stars in the graph (and stars in general) is the same. If this condition is satisfied, all we have to do is measure the luminosity of a star, and we can determine its mass.

But is this ratio the same for all stars? Let's begin with our sun. It's important because astronomers use it as a standard; in other words they take the ratio of its mass to the amount of light it emits to be 1. Since we'll be referring to this ratio frequently

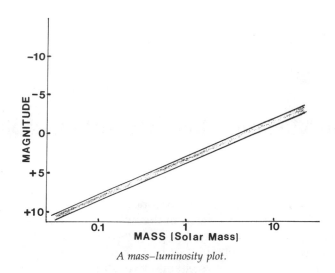

A mass–luminosity plot.

throughout the book, let's give it a name: we'll call it the ML ratio. With the sun as a standard we can compare all stars to it; furthermore, we can even compare galaxies and clusters of galaxies to it.

Now back to our question: Do all stars have a constant ML ratio? If we look at the stars around us that are about the same size and brightness as the sun we find that they do. When we go to much brighter stars, and much dimmer ones, though, we find this is no longer the case. Sirius, for example, which is a blue-white giant with a luminosity 22 times that of the sun, has an ML ratio of only 0.1. This means that it gives out more light per unit mass than the sun; in effect it's a more efficient source. The tiny dwarf companion of Sirius, on the other hand, has an ML ratio of 450. This means that for its mass it's hardly giving off any light. It is, in fact, so dim that we can barely see it in the glare of Sirius—even though it is close (only 8 light-years away) and roughly as massive as the sun. For its huge ML ratio it almost seems as if some light is missing. Obviously, if we included many stars of this type in our plot they would throw things off.

Most stars, however, are not at these extremes. If we take a representative sample of stars around us there will usually be a few more luminous than our sun, and a few dimmer, and things will average themselves out. After all, this is what we would expect since our sun is an average star. It is possible, though, that we could select, say, 20 stars around us and get an ML ratio of 10. This would tell us that we had included a lot of dim, yet massive stars. In other words, for the mass we had there would appear to be little light. On the other hand, if we didn't know better, it might seem that for the light we had there should be a lot more mass. In effect, some mass was missing.

The first person to notice that there was a problem of this sort in the stars around us was Jan Oort of Holland. Born in 1900, his first interest in astronomy came from the books of Nicolas Flammarion and Jules Verne. "I particularly liked Verne's *Voyage to the Moon*," he said. "I also found Hector Servadac's

Jan Oort.

description of a voyage through the solar system on a comet to be delightful."

His greatest inspiration for astronomy, however, came from the famous Dutch astronomer Jacobus Kapteyn. By the time he was in high school he had already developed a strong interest in astronomy, and upon graduation he went to the University of Groningen mostly because Kapteyn was there. At this stage, however, he had not yet decided between physics and astronomy as a career. Kapteyn's lectures soon convinced him: it was astronomy that he was really interested in. "I was particularly impressed by the way he taught elementary celestial mechanics," said Oort.

Oort was fascinated by the dynamics of galactic motion almost from the beginning of his study of astronomy. And this was perhaps expected: the major astronomical project at the University of Groningen while he was there was the study of the dynamics of the Milky Way. He read everything he could find on the subject: books by Newcomb, Jeans, and Eddington. And when he came to do his thesis project, as expected, he chose the dynamics of the motion of stars in the Milky Way. It was the beginning of a lifelong project in which both the size and the rotational motion of our galaxy were determined.

Oort eventually teamed up with the Swedish theorist Bertil Lindblad. Together they showed that our galaxy did not rotate like a solid disk; in the vicinity of the sun it rotated like the planets of the solar system, with the stars closer to the center moving faster than the sun, and those farther out moving slower. Their early papers were not only important in themselves; they also encouraged others to join in the research.

But Oort has not been involved only with galaxies. He has also done a considerable amount of research on comets, and is well known for his explanation of short- and long-period comets. To explain the long-period ones he hypothesized that the sun is surrounded by a cloud of billions of comets. This cloud, now called the Oort Cloud, is out about 1 light-year from the sun. The long-period comets—those with periods of 50,000 years or longer—come from it, according to Oort.

His love of research, and also his love of life, is evident in a paragraph he wrote recently in a brief biography. He was 81 at the time. It reads, "A Dutch historian recently wrote a book on her life called, 'Looking Back in Astonishment.' My own life has been, and still is, one of marvelling about what lies ahead. If I were to write a book on it I would rather call it, 'Looking Ahead in Wonder.'"

In 1932 Oort decided to measure the mass of a given volume of our galaxy (a cylinder around us, perpendicular to the disk's plane). Realizing that stars wander slightly as they orbit our galaxy he devised an ingenious way of determining the mass of this volume. Stars are occasionally carried high above or below the central plane of the disk, and the farther they get from the center, the greater the force pulling them back. Eventually they begin to move back, but by the time they get to the center they have considerable speed and pass right through it. Their overall motion is therefore a slow oscillation through the center of the disk. Oort found that if he could measure the velocities of stars

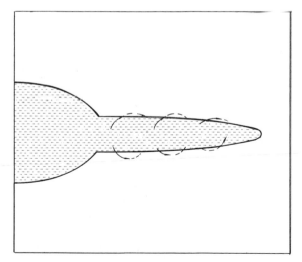

Orbits of stars that stray above and below the disk of the Milky Way.

as a function of their distance above or below the center he could determine the net force acting on them. And this, in turn, would give the mass of the region.

To his surprise, when he performed the experiment he found that there appeared to be considerably more mass in our region of the galaxy than the sum of the masses of the individual stars indicated. This extra mass eventually became known as the "Oort Limit."

I asked Oort what led him to this project. He replied, "The investigation of the force exerted by the galaxy in the direction perpendicular to the galactic plane was a natural extension of what Kapteyn and van Rhijn had done previously. By 1930 the data were extensive and . . . reliable. But there was still considerable uncertainty, so I have never felt very convinced of what you call my 'discovery.'"

Astronomers puzzled over Oort's result but there was little interest at first. By the mid-1930s, however, the structure of the universe was better understood. Hubble had shown that the universe was dotted with galaxies. In fact, as far as telescopes could penetrate there were galaxies. But they weren't uniformly distributed; Hubble showed that they tended to cluster. Indeed, single galaxies were rare. We now know that most galaxies are in clusters of from a few up to thousands. Our own galaxy is in a cluster of about 30 galaxies; it is the second largest of the group, usually called the Local Group. Other clusters contain thousands of members. The Virgo cluster, for example, contains about 2500 galaxies. And another in Coma Berenices called the Coma cluster may contain as many as 10,000.

Fritz Zwicky of Caltech and Mt. Wilson Observatory in California was studying the Coma cluster in 1933 when he discovered, as Oort did in studying our galaxy, that there appeared to be a problem in relation to its mass. Two different methods of determining it gave two different answers. And the discrepancy in this case was much larger (200 times) than in Oort's case.

Zwicky was born in Bulgaria in 1898. He studied at the

Fritz Zwicky.

Federal Institute of Technology in Zurich, where he obtained his B.S. in 1920 and his Ph.D. in 1922. In 1925 he came to the United States. After working for a few years in various fields of physics—jet propulsion, physics of crystals, liquids, and gases—Zwicky eventually switched to astronomy. The way he got into astronomy is interesting. He was having dinner at the home of Robert Millikan, the president of Caltech, when the conversation turned to what distinguishes a "great" scientist. "Any truly great scientist can change fields and make a name for himself within five years," said Zwicky. Millikan quickly challenged him. "All right," he said, "Caltech is going to open a department of astronomy. I'll make you a professor of astronomy and astrophysics, and you have five years to make a name for yourself in the field." Zwicky accepted the offer, and most will agree that he met Millikan's challenge.

To say that Zwicky was prolific may be an understatement. In addition to speaking seven languages, he made contributions to the study of supernovae, galaxies, distribution of galaxies, neutron stars and he invented a method of study he called the "morphological method." One of his most important contributions came shortly after Hubble showed that the Andromeda Nebula was a separate "island universe" of stars—a galaxy at a distance of 800,000 light-years (later corrected to 2 million light-years). Zwicky remembered that a bright star had flared up in the Andromeda galaxy in 1885. It was so bright that it rivaled the nebula itself. If Andromeda was 800,000 light-years away the exploding star could not have been an ordinary nova; it was far too brilliant. He called it a supernova, and, along with Walter Baade, began looking for similar objects in other galaxies, and he soon found several. Later, he hypothesized (correctly) that small dense stars composed mostly of neutrons would be found in such explosions.

For all his contributions to science, Zwicky was a complicated personality. He was a sincere humanitarian, spending many years helping war orphans and working to replace war-torn libraries in Europe. Yet he had a pessimistic outlook on life, and felt that he was being persecuted by many of his colleagues. In fact, he had little to do with most of them. Furthermore, he actually attacked some of them in his book *Morphological Astronomy*. He was, indeed, given little access to the larger telescopes at Mt. Wilson and Palomar (the 100- and 200-inch reflectors). Most of his work was done on an 18-inch telescope.

Some insight into his feelings comes from Arp's book *Quasars, Redshifts and Controversies*. Arp had just discovered a bridge between two objects with different redshifts on a photograph. He writes, "I opened the door into the hallway to see whether someone might be there to share this great moment. Fritz Zwicky was walking by and I asked him with some trepidation because he always tended to be caustic about other people's work. He looked at the plate for a long time and finally declared, 'I am glad you discovered that and not one of those other bastards.'"

Arp went on to say, ". . . in my opinion Zwicky was one of the most creative, hard-working and renowned astronomers who worked at Caltech—but he resented being reduced in observatory time and excluded from councils and committees."

In 1933 Zwicky turned his attention to the Coma cluster in the constellation Coma Berenices. He wanted to determine its mass. Applying what we refer to as a "virial technique" he was able to calculate the overall mass of the cluster. He then calculated the individual masses of the galaxies and added them together. Comparing this to the overall mass he found a considerable discrepancy; the overall cluster weighed a lot more than the sum of the weights of the individual galaxies. The ML ratio was approximately 500. This meant that there was a lot of mass unaccounted for, or else the cluster was in the process of flying apart. Since there was no visual evidence it was separating, Zwicky assumed there was a large amount of "dark matter" in the cluster—nonluminous gas, dust, and so on—matter we couldn't see.

Shortly after Zwicky published his paper, Sinclair Smith of Mt. Wilson did a similar analysis on the Virgo cluster in the constellation Virgo. And he got the same result: there had to be a lot of dark matter in the cluster. In concluding his paper he wrote, "It is possible that both values [overall mass and sum of individual masses] are essentially correct, the difference representing internebular material, either uniformly distributed or in the form of great clouds of low luminosity surrounding the nebula."

In 1937 Zwicky wrote a second paper on the Coma cluster. Further checks had convinced him that the ML ratio was indeed close to 500. He wrote, "The discrepancy is so great that a further analysis of the problem is in order."

But very little in the way of "further analysis" was done for many years. Most astronomers were not terribly interested. It

A cluster of galaxies (in the constellation Hydra). Zwicky showed that much of the mass in clusters is unaccounted for. (Courtesy National Optical Astronomy Observatories.)

seemed like the sort of problem that would just go away when better instrumentation and so on came along. But it didn't. The problem remained through the 1940s and into the 1950s. Then in the mid-1950s the Soviet astronomer V. A. Ambartsumian suggested that no mass was missing; the cluster just had excess energy. This meant that clusters were flying apart, and could last no longer than another 100–1000 million years. Since it was also known that there was a discrepancy in individual galaxies he suggested that the streams of luminous material (jets) observed in some of the radio galaxies were a result of this excess energy. He even went as far as suggesting that arms of galaxies might be a result of it. He did not, however, explain where the energy came from.

Considerable controversy resulted, and because of it a conference was called in 1961 at Santa Barbara, California. Ambartsumian was a central figure. He discussed his ideas, but there were many at the conference who were unwilling to accept them. They preferred to believe that there was a large amount of mass that we could not see or detect. Most believed it was in the form of objects of low luminosity or intergalactic gas. The difficulty with this explanation, of course, was that 99 percent of the mass of the cluster was in this unseen form, and this was still difficult to accept at this point.

A serious problem with Ambartsumian's hypothesis surfaced when astronomers began calculating the "expansion age" of clusters (time since they began flying apart) and comparing it to the age of the universe. Clusters appeared to have an expansion age of 10^9 years, which was considerably less than the accepted age of the universe. This meant that they began to expand well after the universe formed, which did not seem logical.

Thornton Page, who had just completed an extensive study of binary galactic systems, indicated that there was a further complication. He stated that he had found relatively small ML ratios (1–2) for binaries that consisted of spiral galaxies, but large ones (40–80) for binaries that consisted of elliptical galaxies.

Oddly enough, no further work was done on binary systems until 1974 when Edwin Turner, a graduate student at Hale Observatory, began a study. He observed over 150 pairs and concluded that Page's results were in error. He found ML ratios of around 50 for spiral pairs and around 100 for elliptical pairs. The difference between the two results was no doubt due to the more modern equipment and computers that Turner used.

Was there any other way around the apparent mass discrepancy in clusters? It was possible that the intergalactic forces were not entirely gravitational. And indeed this suggestion was made. But most scientists did not feel comfortable with this solution; it introduced "new physics" into the problem.

By the early 1970s Ambartsumian's ideas had generally died away and astronomers were beginning to accept the idea that there was a considerable amount of "dark matter" in clusters of galaxies. Then came an important discovery in relation to individual galaxies.

CHAPTER 4

Dark Matter in Galaxies?

By the mid-1970s it was clear that something was definitely wrong. Not only was there matter in clusters of galaxies that we couldn't see, but there were also indications that we weren't seeing some of the matter of individual galaxies. Our own galaxy was no exception.

To understand the problem more fully it is best to look more closely at galaxies. Simply, a galaxy is a group of a few hundred billion stars. In reality, though, there is much more to it than this. As seen through a telescope they may appear uninspiring, but a long-exposure photograph reveals them to be among the most beautiful objects in the universe. Most beautiful are the spirals, which look like giant pinwheels of glowing gas and luminous blue-white stars. A closer look reveals that they appear to have a dense spherical core with two or three dangling arms wrapped around it. But as we will see appearance is deceptive.

They are indeed like pinwheels in that they spin slowly about their axis. You could watch one for years, though, and never see the slightest movement. It is revealed only through the use of the spectrograph. In some cases the arms trail loosely, in others they are tightly wound. We refer to the tightly wound ones as "a" type, and the loosely wound ones as "c" type ("b" types are medium tightly wound).

Besides spirals there are elliptical galaxies. They have no arms, and, as their name implies, they are elliptical, or egg-shaped. They are also unlike spirals in that the stars in them do

A spiral galaxy in the constellation Triangulum. (Courtesy National Optical Astronomy Observatories.)

not all rotate in the same direction. Furthermore, they have little gas or dust.

Where did galaxies come from? How did they form? According to the big bang theory, about 18 billion years ago the universe was a rapidly expanding homogeneous gas cloud. This cloud eventually broke up into smaller clouds and each of these smaller clouds collapsed in on itself creating, in time, a galaxy.

The overall features of galaxy formation are generally agreed upon, but many questions remain unanswered. For example: how did the spirals begin to spin? Certainly, when the collapse began the particles were in random orbits. As the collapse proceeded, though, the randomness somehow smoothed out and most of the particles ended up moving in the same direction. Another problem is related to how the different types of galaxies formed, and how the various types evolved. Astronomers are not yet able to answer these questions in a completely satisfactory way. Nevertheless, considerable progress has been made.

One of the galaxies that formed was, of course, our Milky Way. It is a medium tightly wound spiral (type "b") that contains about 200 billion stars. Like all spirals it has an inner bulge, which is cut off from our view by gas and dust. Its center is in the direction of the constellation Sagittarius. If you went out and looked in this direction on a dark night you would see a large number of stars, but in reality you are seeing only a small fraction of those that are there.

We are situated about three-fifths of the way out from the center in an arm called the Orion arm (so called because much of the material of this arm is in the direction of the constellation Orion). The diameter of the Milky Way is approximately 100,000 light-years, so we are at a distance of approximately 30,000 light-years from its center. Closer to the center is another arm called the Sagittarius arm. The arm beyond us is called the Perseus arm.

Although it was clear that there was a problem in relation to the mass of both clusters of galaxies and our own galaxy, it had not drawn a lot of attention, even as late as 1970. The mass discrepancy was large in the case of clusters of galaxies, but it was relatively small for individual galaxies, and most astronomers were convinced that it was just a bookkeeping error in this case. They were sure there were dim stars, planets and interstellar matter that we were not seeing, and that the problem would be cleared up when larger telescopes and better instrumentation were available. For the most part the problem was just ignored.

INSTABILITY

Then in 1973 Jeremiah Ostriker and Jim Peebles of Princeton University published a paper that the scientific community could not ignore. They showed that if there wasn't considerably more matter in our galaxy than we see, it had to be unstable. In fact, it couldn't last long if it was really what it seemed to be. Yet it wasn't unstable, and it certainly existed.

Born and raised in New York City, Ostriker got his undergraduate degree from Harvard and his Ph.D. from the University of Chicago. "I was always interested in science when I was young," he said. "Living in New York, though, I don't think I could see the stars very well. So I wasn't interested in them in the way an amateur astronomer is." Nevertheless, his fascination with astronomy and physics grew. "I read books by Eddington and Jeans. They impressed me with their power of pure logical thought . . . how you could derive things just by thinking. I thought maybe that's the kind of thing I would like to do. It stayed in the back of my mind."

But when he went to college he took a couple of courses in astronomy that changed his mind for a while. "They were terrible," he said. Then, after receiving his degree in physics, he began to think about what he wanted to do with the rest of his

Jeremiah Ostriker. (Courtesy John W. H. Simpson.)

life. "I said to myself: thirty years from now I'll be solving some kinds of equations because I'm good at solving equations, and I could be working for some toothpaste company trying to put the toothpaste in the tube more efficiently. Or I could be using the same equations to figure out the interior of a star, or the origin of the universe. Which would be more fun?" He paused briefly. "That convinced me, so I went to the University of Chicago because Chandrasekhar was there. He was the world's greatest astrophysicist and I wanted to work under him."

Ostriker did his thesis on the stability of rotating stars. His problem was to find the stability limits as stars rotated faster and faster. How fast, in fact, could they spin without becoming unstable? Surprisingly, he found that if they became distorted even by a moderate amount as a result of spin, they would become unstable and might break up.

When Ostriker graduated he went to Princeton University

where he continued his work on the stability of rotating systems. It soon became clear to him that his method could be applied not only to stars, but also to galaxies. But his background in cosmology at the time was insufficient to tackle the problem alone. Fortunately, in the next building was one of the world's leading cosmologists, Jim Peebles. "I thought my work would interest him," said Ostriker, "so I went to him with the idea of applying it to galaxies and he was intrigued by it."

Born in Winnipeg, Manitoba, in 1935, Peebles got his B.S. at the University of Manitoba in physics. Upon graduation in 1958 he went to Princeton University where he worked under Robert Dicke. "Dicke got me interested in cosmology," said Peebles. "Before I met him I had not thought much about it." And his collaboration with Dicke soon paid off. Within a few years Pee-

James Peebles.

bles was associated with one of the most important discoveries in astronomy—the detection of the cosmic background radiation.

At the time Ostriker visited him, Peebles was studying galaxy formation and had been worrying about the origin of the rotation of galaxies. Peebles described his problem to me. "In the gravitational stability picture that was popular then and still is, there is a puzzle because . . . galaxies are supposed to arise from the gravitational growth of small mass irregularities. But when these irregularities grew they would not necessarily have much spin to begin with. You had to ask: Where did the spin come from? I speculated that it came from tidal torques—the pulling of one protogalaxy [early form of a galaxy] on its neighbors. I computed the amount of these tidal torques and observed that it was reasonable for elliptical galaxies, but a little shy for spirals."

When Ostriker told him about his work on rotating stars Peebles realized that it was closely related to his work on rotating galaxies. Ostriker told Peebles that he was convinced that spinning, disk-shaped objects like spirals should be unstable. "That pleased me," said Peebles, "because it relieved the pressure on my theory to produce spin in spirals." Peebles had written a computer program earlier in relation to a different problem that he was able to adapt to this one. I asked him about it. "It was a naive program, none of the sophistication of today," he said. "We used it to compute the gravitational accelerations . . . and sum the forces due to all the other particles. There are much more efficient programs now, but those were the early days. The program was one I had developed for studying the clustering of galaxies in an expanding universe. It was easy enough to turn off the expansion of the universe and put in a lump of material rather than a homogeneous distribution."

The program verified that spiral galaxies were, indeed, unstable. "We were convinced that they would be unstable before we set up the program," said Peebles, "so we weren't too surprised. No . . . this wasn't a serendipitous discovery." Ostriker's reaction: "I guess I was extremely pleased but not surprised because it confirmed what I had known. In my work on rotating stars I found that whenever they were mainly supported by rotation they were unstable. You could make a general physical argument that showed that they ought to be but you couldn't prove it would also hold for a system like a galaxy. If you gave a star a lot of spin it would turn itself into a bar or a binary. So it made sense that a spiral galaxy would also be unstable."

They found that after a single rotation their disklike system of particles (representing a spiral galaxy) would start to become bar-shaped. And if the spin was very high they would literally fly apart and become a binary system. This was exceedingly strange in that we know that our galaxy is stable. It doesn't look like a bar—and it's certainly not flying apart. There had to be something "extra" in our galaxy holding it together. But what? Ostriker and Peebles tried various things to stabilize it. The only one that seemed to work was to assume that our galaxy was

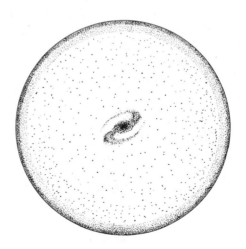

Dark matter halo around a galaxy.

immersed in a halo. Adding this to the computer program they found that it stabilized the galaxy as long as there was at least as much mass in the halo as there was in the galaxy itself. This was surprising because there was little evidence for such a massive halo. I asked Ostriker how he was led to postulate a halo. "There's no reason a . . . [halo] should form," he said. "So if you are going to have a halo—you have to invent it. It makes most sense to have one that is more or less round. And, in fact, we have one already in our galaxy—the spheroid at the center. So basically, just making an extension of that seemed to be the thing to do."

Their paper, titled "A Numerical Study of the Stability of Flattened Galaxies: Or, Can Cold Galaxies Survive?" was published in the December 1973 issue of *The Astrophysical Journal*. A year later they (along with A. Yahil) published a second paper on the same subject. In it they presented evidence that galaxies may also have halos that extend well beyond the outer visible limits of the galaxy. The mass in this halo might, they said, be as great as ten times the observed mass of the galaxy. This would drastically increase the overall mass of galaxies. But what form was the mass in? According to Ostriker and Peebles it was most likely in the form of small, dim stars, or possibly burnt-out stars.

With the publication of Ostriker and Peebles's paper the dark matter problem entered a new era. Not only was there considerable evidence for dark matter in both clusters of galaxies and individual galaxies, but they had shown that if there wasn't considerably more matter in galaxies than we see they wouldn't exist. Astronomers were confused. The dark matter had to be there—why couldn't we see it?

RUBIN AND FORD

Even before Ostriker and Peebles had made their discovery, another similar one had been made by Vera Rubin and Kent

Ford of Carnegie Institution. They discovered that one of our nearest intergalactic neighbors, the Andromeda galaxy (usually designated by its Messier number, M31), rotated in a strange way. But at this stage they were not ready to suggest that it was a general phenomenon. It could be peculiar to M31.

Their project was made possible because of a special image intensifying device (called an image tube) that had just been developed at R.C.A with support from Carnegie. Kent Ford Jr. worked on the tube while still a graduate student at the University of Virginia. Upon graduation in 1957 he became a full-time staff member, and continued to work on image tubes and the spectrographs that were used with them.

Vera Rubin was born in Philadelphia. "I became interested in astronomy at about age 12 or 13 by looking at the stars through my bedroom window," she said. Upon graduation from high school she went to Vassar to study astronomy, selecting Vassar "because most other colleges that taught astronomy did not accept women." After graduating from Vassar she married a Cornell graduate student and joined him at Cornell where she got a master's degree. She then moved to Washington, D.C. and in 1954 obtained her Ph.D. from Georgetown University. She stayed there for 10 years, first as a research assistant, then as an assistant professor. It was here that she first became interested in the rotation of galaxies, but Georgetown was isolated from the astronomical community so in the early 1960s she traveled to the University of California at San Diego where she worked with Geoffrey and Margaret Burbidge, a well-known husband–wife team who were active in galaxy and quasar research. For a year she took spectra of quasars, but the frantic pace of the research soon got to her. Therefore, upon returning to Washington, she asked the Carnegie Institution for a job. Many years earlier she had decided that was where she wanted to work.

At Carnegie she teamed up with Kent Ford. "We worked well together," she said. "Kent built the exquisite equipment; we both observed and spent a lot of time crushing heads at the

Vera Rubin. (Courtesy Mark Godfrey.)

telescope, then I measured and reduced the spectra while he went off to build more equipment."

Rubin and Ford's first major project was a study of the rotation of M31. The ideal candidate for such a study is a galaxy that is inclined (the Doppler shift between the two sides is high in this case). And M31 is inclined at 77 degrees, and is also one of our closest galaxies (2.2 million light-years away) so it is relatively easy to study.

Because of its tilt, one side of M31 is approaching us, and the other edge is receding from us. This means that the spectral lines from the approaching side will be blueshifted and those from the receding side will be redshifted. Rubin and Ford took data at many points along the galaxy, both on the receding and the approaching sides. Considerable preparation and several hours at the telescope were required for each point, but by 1969 they had observed a total of 67 regions. And their results left little doubt: M31 was rotating in a strange way. The relative velocities of the stars and gas clouds out near the edge of the galaxy were much higher than expected. If the outer regions of M31 rotated like the solar system—as was expected—their velocity should have been low. This was puzzling because it could happen only if there was considerable mass in this region of the galaxy, and observational evidence indicated there was little light there. The center of the galaxy was the brightest and it seemed reasonable to assume this is where most of the mass was.

I asked Rubin about her reaction to the result. "My reaction was delight that we had discovered something we had not expected. This is the greatest reward that you can ask from any research program," she said. "I had been interested in motions within galaxies since my student days . . . in order to learn about the distribution of mass in the universe. But I had to wait until telescopes and equipment were advanced enough to

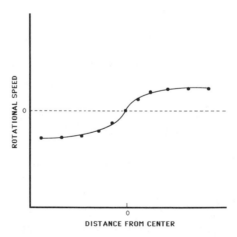

A plot of rotational speed versus distance from the center for the Andromeda galaxy.

permit me to observe the faint outer regions of galaxies. I learned slowly, so there was no instantaneous recognition that rotation curves were flat. It took many many nights before I knew what we had."

The results on Andromeda were puzzling but by 1970 the work was complete and Rubin went on to another project. Ford continued working on the image tubes in an effort to improve them. Oddly enough, even before Rubin and Ford had completed their research in the optical region of M31, radio astronomers were busy looking at the radio emissions from the same galaxy. Astronomers at Green Bank, West Virginia had begun to study the hydrogen emissions (21-cm radiation) from M31 and other nearby galaxies. Among them was Morton Roberts of Harvard. And not surprisingly he showed, as Rubin and Ford did, that the velocities of the gas clouds in the outer regions of M31 were much higher than expected.

By 1975 image tubes had been significantly improved and Rubin and Ford were ready for another, more extended look at the problem. Again they planned to look at rotation curves of galaxies and use them to determine the distribution of mass within the galaxies. This time they selected not one, but 60 different spiral galaxies, ranging through types a, b, and c, and also including both intrinsically bright and dim galaxies.

Before I talk about their study let's take a moment to consider what we would expect a rotation curve to look like. A typical spiral galaxy has a dense spherical center with long dangling arms wound around it. Kepler's laws tell us that if the mass is indicated by the brightness the outer stars in this system should rotate like the planets in our solar system. In the outer regions we would therefore expect the motion to fall, i.e., be Keplerian. What is important in the case of a galaxy is the mass of the material (stars and so on) between the star we are considering and the center of the galaxy; the star's orbit is determined by this mass. In the solar system we have the same situation, but literally all the mass is in the sun. If you tried to determine how much mass was inside, say, the Earth's orbit, you would just get the sun's mass—the mass of Mercury and

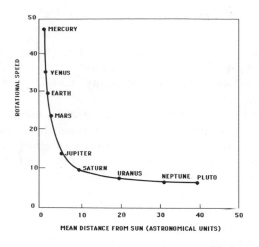

A plot of speed (vertical axis) versus distance from the sun for the planets.

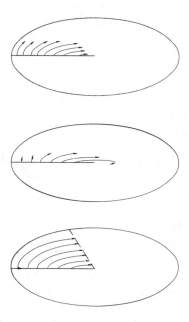

Top: The way a galaxy rotates.
Middle: The way the planets of the solar system rotate.
Bottom: Solid body rotation.

Venus are negligible. But in the case of a galaxy, the mass interior to a given star is spread out over many stars. With this in mind, then, let's look at what we would expect. We'll plot the total mass inside a particular radius against distance outward in a spiral galaxy. From the center through the central bulge we would expect the mass to increase linearly, and quite sharply, as shown in the graph. Once we get to its edge, the density of mass drops off sharply so we would expect the slope to be less steep, as shown. And finally, when we were beyond the outermost visible region of the galaxy we would expect the curve to remain level. A plot of rotational velocity versus distance would be similar to this but should fall off once we were in the outer regions (with most of the mass on the inside of the orbit), just as it does in the case of the solar system.

Rubin and Ford had already shown that this was not the case for M31. But did it apply to all galaxies? Equipped with the new image tubes they, along with two colleagues, Norbert Thonnard and David Burstein, began taking data using the huge 4-meter telescopes at Kitt Peak, Arizona, and at Cerro Tololo in Chile. This time, though, there was a difference in the way they took the data. Previously, the image of the Andromeda galaxy had been so large that they could only take the spectrum of a small section of it at a time. The 60 spirals selected by Rubin and Ford, however, were much more distant, and therefore presented a smaller image in the sky. They were therefore able to place the slit of the spectroscope across the entire visible extent of the galaxy—along its axis. In this way they could get all the red- and blueshifts they needed at once. In other words, a single exposure gave the position of the spectral line for all distances from the center of the galaxy. Furthermore, with other advances, each exposure now took only about two or three hours.

The first thing they noticed was that there was a difference in rotational speed depending on what type of spiral they were observing. The tightly wound spirals ("a" type) rotated the

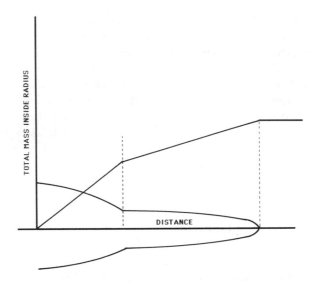

A plot of how total mass should vary with distance outward in a galaxy.

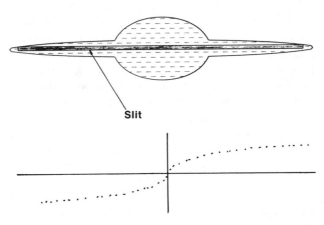

Upper: Spectroscope slit shown along length of galaxy.
Lower: A plot of rotational speed for various distances along the galaxy.

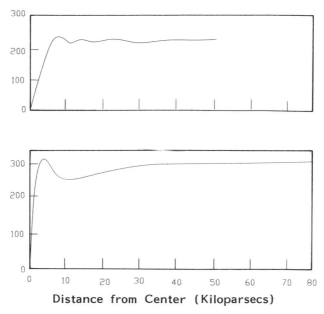

Distance from Center (Kiloparsecs)

Velocity curves of galaxies. Vertical axis is rotational speed.

fastest, the loosely wound ones, the slowest. But this wasn't what surprised them the most. Most amazing was the shape of literally every curve: they were all similar, and the velocity of the gas clouds in the outer regions of all 60 galaxies was high. It did not drop off as would be expected if the motion was Keplerian (and the light gave an indication of the mass present). This meant that there had to be a tremendous amount of matter in the outer region of all spiral galaxies—matter that we couldn't see directly.

It's not that the density didn't fall off; it did, but not nearly as fast as the light intensity. Thus, just as Ostriker and Peebles had shown earlier using a computer program, Rubin and Ford and their colleagues showed that galaxies had to have a massive halo.

Further verification came from others. Albert Bosma of Holland measured the same curves for 25 nearby galaxies using the Westerbork radio telescope. His results (a Ph.D. thesis project at the University of Groningen) were published in 1978. And, like Rubin and Ford's curves, Bosma's were flat not only out to the edge of the visible light, but also slightly beyond it.

What was amazing about the result was that it meant that most of the mass of our galaxy and others—about 90 percent—is not detectable in the visible light we see. It is in a gigantic halo that surrounds the galaxy, extending out considerably farther than its visible edge.

It was such an important result that further verification was needed. Is there any other way we can check it? One way would be to look at objects beyond the edge of our galaxy and see how our gravitational pull affects them. There are many objects under the gravitational influence of our galaxy: globular clusters, satellite galaxies such as the Magellanic Clouds, and halo stars. Ostriker and Peebles were two of the first to consider this. "We looked at the satellites as you went further and further from our galaxy and they indicated more and more mass," said Ostriker. "We didn't do the observations . . . just collated all the observations that existed at that time. People had found these things but they had blinders on and didn't believe them. We just put it all together."

Further checks, some of them using objects at great distances from our galaxy, were made by Tadayuki Murai and Mitsuaki Fujimoto of Nagoya University, C. Lin of Lick Observatory, Donald Linden-Bell of Cambridge University, and Jaan Einasto of the USSR. Soon there was no doubt: even objects at considerable distances from our galaxy were affected by its unseen mass.

A similar check of the globular clusters of M31 was made by Stephen Shectman, Leonard Searle, and Peter Stetson of Mt. Wilson and Las Companas observatories. Surprisingly, they found that, on the average, the globular clusters of M31 rotated around it considerably more slowly than the galaxy itself rotated. Furthermore, the variation in velocity was large, and nearly a third of those observed rotated in the reverse direction. This seemed to be in conflict with results on our galaxy. Why is it different? We still don't know.

But if the halo really is there, what is it composed of? If it were made of stars of average brightness we would have photographed them long ago. Also, it can't be clouds of hydrogen gas; we can easily detect the radio emission from them. And it can't be dust; dust blocks light and we would soon notice it. The best bet seems to be extremely dim stars, or perhaps dead stars—stars that have gone through their life cycle and no longer have a thermonuclear furnace.

Also, how and why did these objects form? As I mentioned earlier, we believe that our galaxy was created in the collapse of a giant pregalactic cloud. Did the halo form before the disk formed, or after? In an effort to answer these questions Allan Sandage of Hale Observatory made a study of a large sample of the stars around our galaxy, but outside it (excluding globular clusters)—the halo stars. He found that there was a strong correlation between orbital velocity and heavy-element content. This seemed strange. We know that stars do not gain heavy elements until they are quite old. And therefore only stars that formed late in the evolution of the galaxy could be expected to be rich in heavy elements. If we found halo stars, or globular clusters, that were rich in heavy elements and moved rapidly in their orbits, they would reinforce our hypothesis that galaxies formed in the collapse of a spinning gas cloud. And this is, indeed, what Sandage found.

In the same year that Sandage published his paper (1984) Robert Zinn of Yale University reported that heavy-element-rich globular clusters of our galaxy rotated at a relatively fast rate, while older, heavy-element-poor globular clusters have virtually no spin, further verifying the hypothesis.

So far I've talked only about spiral galaxies. And there is a reason for this: it is much more difficult to measure the velocity curves of elliptical galaxies. The stars in them do not orbit in the same direction, as they do in spirals. They are, for the most part, like bees around a hive; each star has its own distinct orbit, different from all others. Furthermore, there's another problem with ellipticals: they have no hydrogen clouds. And, as we saw earlier, velocity curves of spirals are usually made using hydrogen clouds.

The first attempt to obtain velocity curves of ellipticals was made in 1975, but with little success. Astronomers found very little overall rotation of the ellipticals they checked. This, of course, didn't come as a surprise, and fortunately it didn't discourage others from trying. Paul Schecter of the University of Arizona, James Gunn of Caltech, and Garth Illingworth of Kitt Peak Observatory decided to take a different approach. Instead of trying to get the rotation curves directly, they analyzed the thickness of the spectral lines. If the line is broad there would be a large variation in orbital velocities among stars in the galaxy. They encountered considerable difficulty but finally got some results. They found that the flattened ellipticals had the greatest tendency to have organized rotation. This helped to simplify things.

The overall result of these studies is that ellipticals are like spirals in that they also have considerable unseen mass. The evidence is not as strong, but where measurements have been taken the velocity is high in the outer regions of the galaxy.

Evidence has also been obtained for the existence of an enormous invisible halo around the elliptical galaxy M87. It is well known that M87 is a strong source of radio and X rays, and it was therefore a prime target for study when Riccardo Giacconi and his colleagues launched the first X ray satellite into space in 1978. The results were startling: M87 had a halo of X rays around it that was 50 times larger than its visible radius. Such a large halo would only be held to the galaxy if it was 50 times as massive as we previously believed it to be (based on the light we see).

Do all elliptical galaxies have halos such as this? Certainly, we don't know. So far there is no indication that they do, and it's obvious that much more work is needed.

The problem with M87, and all galaxies for that matter, is so difficult to comprehend that some astronomers have suggested the only way out is to modify Newton's law of gravity over large distances. But again, few accept this.

CHAPTER 5

Dark Matter in the Universe?

We have seen that there is dark matter both in clusters of galaxies and in individual galaxies. But galaxies are part of the overall universe. Does this mean there is a dark matter problem there also? There is, indeed, an enigma in relation to the universe, but it may or may not be related to the dark matter of galaxies. Some of the universe's matter appears to be missing, so in this case we may have a real missing mass problem. In the case of galaxies and clusters the mass isn't missing; we have strong evidence that it's there. We just don't know what form it takes.

To understand things more fully it is best to go back to the beginning—the beginning of the universe. We talked about it briefly in an earlier chapter. As we saw there the most satisfactory theory of the origin and evolution of the universe is the big bang theory, a theory in which the universe is assumed to have begun about 18 billion years ago in a tremendous explosion. The expansion caused by this explosion is still going on; we have ample evidence of it in the form of redshifted spectral lines. According to the Doppler interpretation of this shift all galaxies appear to be moving away from us, and the farther they are away from us the faster they are going. It's not that they are moving away just from us; they are actually moving away from one another. It's the space between the galaxies that is expanding. Therefore, the universe would appear the same regardless of which galaxy we happened to be on.

But if the universe is expanding, it's natural to ask: What will eventually happen to it? There are only two possibilities: it can continue to expand forever, or it can eventually stop expanding and collapse back on itself. We can easily see why this is so if we compare it to an explosion here on Earth. In a terrestrial explosion debris is blown out in all directions from the blast; some of it goes upward and some out to the sides. A piece that goes directly upward will be flung outward at high speed, but the gravity of Earth will pull back on it. It will move upward until gravity finally stops it and it falls back to Earth. It is possible, though, that it could completely leave the Earth. If its speed was initially greater than what is called the "escape velocity" it would fly off into space never to return. Escape velocity is the speed needed to completely overcome a gravitational field (it depends on the mass that creates the gravitational field). In the case of Earth it is approximately 25,000 miles per hour. If a rocketship has this speed at liftoff, it will not fall back to Earth, nor go into orbit around it, but will fly off into space.

In short, then, our piece of debris will fall back to Earth if its speed is less than the Earth's escape velocity, or it will fly off if it is greater. Let's return, then, to our discussion of the universe. It also began as an explosion, and gravity is also pulling back on it. Before we talk about it in detail, though, let's simplify things a little. One of the basic postulates of modern cosmology is that the universe is homogeneous; in other words, it's the same everywhere. This means that if we select a large sphere of galaxies around us, it will be the same as any other similar sphere in the universe. And since there will be no net external gravitational forces on it as a result of the galaxies around it—a push or a pull on one side will be exactly balanced by a similar one on the other side—anything that happens in the universe will also happen in our sphere. This means that we can look at what happens in our sphere and extrapolate it to the universe.

So, let's see what will happen. There is, of course, a mutual gravitational pull between all the galaxies; in other words, they are all attracting one another. And if this mutual attraction is great enough they will stop moving outward. How great does it have to be? For an answer let's go back to our terrestrial explosion. We saw that if the velocity of the debris was greater than the escape velocity it would fly off. Similarly, in this case if the velocity of the galaxies is greater than the "escape velocity" that results from their total mass, they will separate forever. What this means is that if the overall mass, or density (mass per unit volume), of the universe is over a certain critical amount the outward expansion will be stopped. If it is less it will not.

This critical density, it turns out, is approximately 2×10^{-29} g/cm³. If the universe has an average density greater than this it will eventually stop its outward expansion and collapse back on itself; if this happens it is said to be *closed*. On the other hand, if the average density is less than this, the universe will go on expanding forever. It is then said to be *open*.

The obvious question is then: Is the universe open or closed? To answer this we merely need to measure its average density. But, as it turns out, this is easier said than done.

Before we talk about the difficulties let's take another look at our open and closed universes. Consider the closed one first. If we could make large-scale measurements in it we would find something strange: the space within it would appear to be curved. How can space be curved? It is, indeed, a strange concept; certainly there's no way we could see the curvature directly. We can, however, see the curvature of a two-dimensional surface, say the surface of an orange. The best way, therefore, to think about three-dimensional curved space is to think about what happens in two dimensions, then extrapolate to three dimensions. In the case of the closed universe the space is positively curved, like the surface of a ball. If we start out at two different points, side by side, and attempt to draw two parallel lines, we find that they eventually cross. Similarly, we know that the sum of the three interior angles in a triangle is 180 degrees if the triangle is drawn on a flat surface. If, however, we

The galaxies in this photograph are expanding away from us. If the universe is closed they will eventually move toward Earth, and in time, all such clusters will collapse together. (Courtesy National Optical Astronomy Observatories.)

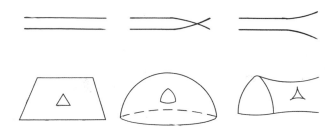

Left: Flat two-dimensional space.
Center: Positively curved two-dimensional space.
Right: Negatively curved two-dimensional space.
The upper diagrams show rays (or lines) that start off parallel in the space.

sum the angles of a triangle that is drawn on the surface of a ball (positively curved) we find they sum to more than 180 degrees. Extrapolating to three dimensions we find a similar situation. Lines that start out parallel, eventually cross, and the sum of the angles in a triangle is greater than 180 degrees. In the same way, if the lines diverge, or the sum of the interior angles is less than 180 degrees, the space is negatively curved. The two-dimensional analogy of negatively curved space is the surface of a saddle.

Incidentally, I should mention that all this relates back to a cosmology we talked about in Chapter 2, namely Friedmann's cosmology. In 1921 A. Friedmann showed that if the universe is positively curved it is closed, and will collapse back on itself. Similarly, if it is negatively curved, it is open and will expand forever. The dividing line between these two cases is the flat universe.

This means that we have two methods of determining whether the universe is open or closed. We can check its average density, or we can attempt to directly observe its curvature. Let's look at density measurements first.

DENSITY TESTS

Suppose we wanted to measure the average density of the universe. What would we do? The most logical thing would be to add up the masses of all the visible galaxies, and divide by the volume that they are in. And, indeed, this was done many years ago for the first time by Edwin Hubble. Since then a considerable amount of work has been done, and there now seems to be no doubt: the amount of visible matter in the universe is far short of the amount needed to close it. The average density calculated using visible matter is about 2×10^{-31} g/cm^3, which is only 1 percent of the critical density.

Does this mean the universe is open? Not necessarily. First of all, the density calculated from galaxies alone is only a lower limit. There is no doubt considerable matter in the universe we can't see. In fact, we saw evidence for this in earlier chapters. But is there enough to close the universe? We don't know for sure yet, but if the universe is, indeed, closed there's a lot of matter we can't account for.

But why would we believe the universe is closed? We will see later that there are theoretical reasons. Aside from this, though, there is what you might call a prejudice. A closed universe is much more satisfying to most astronomers than an open one. But if the universe is closed we do, indeed, have a missing mass problem. Where is all the mass that is needed to close it?

TESTS OF CURVATURE

I mentioned earlier that in addition to density tests there are also direct tests of curvature, and that they can also tell us if the universe is open or closed. Let's look at them. The first is called the *standard candle test*. To understand it let's begin with a galaxy somewhere out in space. As you likely know, it emits light in all directions, and if the space around it is flat there are the same number of rays emanating from it in one direction as there are in

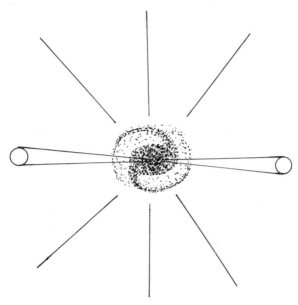

The number of rays emitted in a small cone will be the same regardless of the direction of the cone. But if the space is curved a different number of rays will be emitted in the cone.

another. This means that there are the same number in a small cone in one direction as there are in a similar cone in a different direction.

Now, suppose we distort or curve the space. What happens? Obviously the number of rays within the cone will change. You can, in fact, show that if the space is distorted so it is positively curved, the number of rays in the cone will increase (as compared to flat space). Similarly, if the space is negatively curved the number will decrease. This means that if the universe is positively curved a galaxy will appear brighter than it actually is (or would appear if the space was flat).

How can we use this to test the curvature of the universe? First, we need what is called a standard candle (a reference galaxy). If we can find a large number of such galaxies, and if we

know the brightness and distance of both the nearby ones and
the distant ones, we can check to see if the more distant ones are
over- or underluminous. And we do, indeed, have such a group. They are not all
exactly of the same brightness, but they are close. Astronomers
have noticed that at the center of many clusters are giant elliptical galaxies, and to a first approximation they are all of the same
brightness. M87 in the Virgo cluster is a good example of this
type.

It would seem, then, that all we would have to do is compare a distant one with a nearby one and we would know if the
universe is positively or negatively curved. But, as in all tests of
this type, there are problems—serious ones. First, the ellipticals
are not all exactly the same so we have to measure as many as
possible, then average them. Second, although giant ellipticals
are visible far out into the universe, the galaxies in the associated cluster are not. This means that in the case of the most
distant giant ellipticals we can't be sure we have the right kind.
And third, we have to ask: do elliptical galaxies evolve as they
age? If they do, significant changes will result.

Let's take a moment to look at the problem of evolution of
ellipticals. It is well known that ellipticals do not have any gas or
dust; furthermore, they do not have any massive blue-white giants. Literally all the stars arc red dwarfs. This tells us that star
formation is over. When the galaxy was originally formed there
were, of course, stars of all sizes; blue-white giants were present
along with medium-sized stars like our sun, and small red
dwarfs. We know, though, that all stars go through a life cycle.
Like humans, they are born, live out their lives, and die. And
the most massive ones live the shortest period of time; in fact,
after only a few tens of millions of years they explode as supernovae and dissipate off into space. The debris that is left from
the explosion then goes into the making of new stars, most of
them small, with long lifetimes. But if the original star was massive enough, a few giant stars will form along with smaller stars,
and these new giants will explode again (only the giants ex-

plode). This means that eventually there will be no giants left, only small stars—most of them red dwarfs.

When we look at ellipticals we see that they fit this scenario: they have no gas and dust and are made up of mostly red dwarf stars. But they are also at many different distances from us, and because of this we are seeing them at many different ages. The reason is the finite speed of light: when we look out into space we are actually looking back in time. This means that the farther out an elliptical is, the younger it is. But ellipticals had many blue-white giants when they were young, and supernova explosions were common. This means that, on the average, they were brighter early on than they are today. Therefore, if we look at very distant ellipticals they will appear brighter than they would if no evolution (e.g., supernova explosions) had taken place. But this is the same effect—increased luminosity—we're looking for as an indication of curved space. Obviously, if evolution does occur, this test is in trouble, unless we can somehow account for it.

And there's another problem. Theoretical studies show that giant ellipticals are likely to collide with and absorb any small galaxies around them. They may, in fact, capture and "eat" several during their lifetime. This would also increase their luminosity. Needless to say it also weakens our argument that increased luminosity indicates a positive curvature of space.

Without taking evolutionary effects or cannibalism into account this method indicates a closed universe. Studies show that the average density of the universe is from one to three times the critical density. But something obviously has to be subtracted to allow for the above effects. At the present time, though, we're not sure exactly how much to subtract.

Another test of curvature is known as the *angular diameter test*. It is similar to the above test but related to size, rather than brightness. Assume again that we have a number of galaxies, all of approximately the same size. As in the previous case, astronomers usually select giant ellipticals in the center of clusters. If we compare the diameter of a nearby one with that of a distant one we get a measure of the curvature of the universe. The rea-

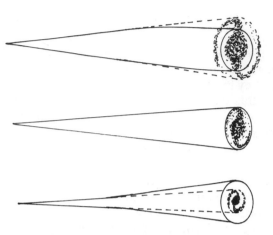

Comparison of a galaxy seen in positively curved space (upper) and negatively curved space (lower).

son is as follows: If the universe is positively curved the galaxy's disk will appear too large (see diagram). Similarly, if the universe is negatively curved the galaxy's disk will appear too small.

These tests are again inconclusive and controversial, mostly because of the effects of evolution. Without corrections for evolution they give a density of 0.4 to 0.8 times the critical density, which indicates an open, or negatively curved, universe.

A third test of curvature is known as the number-count test. Basically, the idea is simple: you merely count the number of galaxies of various magnitudes in a given direction and compare the numbers obtained. The comparison gives you the curvature of the universe because the points that are obtained are plotted on a flat piece of paper. If the universe is positively curved, the plot should also be made on a positively curved surface. When this is done and the curved surface is flattened, an excess of points will appear near the center. This means that if we make a plot of nearby and distant galaxies and there appears to be an excess near us, the universe is positively curved. Similarly, if there appears to be an excess out near the edge of the universe, it is negatively curved.

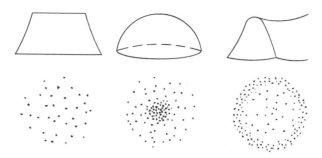

Distribution of galaxies in a flat, a positively curved, and a negatively curved universe.

But again there is considerable uncertainty and controversy. Out of the confusion, though, has come some positive results. The first surveys, taken by optical astronomers, were ambiguous. But then radio astronomers got into the act. They also got ambiguous results but were able to explain why they were ambiguous: the radio sources were evolving. So, although the method has not given much in the way of an estimate of the curvature of the universe, it has told us that there is considerable evolution, and that it must be taken into consideration.

There are several other tests, but I will say little about them, because in each case they are controversial. One is related to the random velocities of galaxies within clusters. Sandage, Tammann, and Hardy showed in 1972 that it was possible (in theory) to get a value for the average density by using these velocities. They published a paper indicating it was 0.04 ± 0.02 times the critical density, again indicating a negatively curved or open universe. Others soon found fault in the result, but I won't go into the details.

NUCLEOSYNTHESIS

By the early 1970s indications were strong that the universe was open but there was still controversy. Then came what seemed to be irrefutable proof. It is related to the generation of the first nuclei in the universe, so let me begin with a brief account of how this occurred. The first nuclei appeared about three minutes after the universe began. Within a few minutes all the light elements of the universe up to helium (along with a small amount of lithium) were produced. First, deuterium, a heavy form of hydrogen was formed in the inelastic ("sticky") collision of a proton and a neutron. Then the deuterium nucleus was hit by a neutron, creating a heavy form of hydrogen called tritium. Tritium, however, is unstable, and most of it soon decayed. Meanwhile, deuterium nuclei were colliding creating helium.

Strangely, though, this was as far as the reactions went. If a helium nucleus is struck by a neutron it forms a nucleus that immediately decays. With no stable nuclei just beyond helium things could go no further—there is, in essence, a gap there that cannot be jumped. (Actually, a small amount of lithium was produced as a result of simultaneous hits.) This period of the history of the universe is referred to as the *nucleosynthesis* era.

What is particularly useful to us is that the creation of deuterium and lithium is related to the average density of the universe at that time. This is, of course, to be expected. If the density of protons and neutrons in the early universe was high, a large number of collisions would have taken place, and most of the deuterium nuclei that were created would have been converted into helium. On the other hand, if the density was low there would have been few collisions and much of the deuterium that was produced would have survived. By measuring the amount of deuterium (and other light nuclei) presently in the universe we can determine what its density was immediately after nucleosynthesis. We can then work forward to determine the average density of the universe today.

What do the measurements tell us? They imply that the density of the universe at the present time is only about 10 percent of the critical density. This might seem like it is the final word.

But it turns out that there are several loopholes. First, it pertains only to a class of particles called baryons (the heavy particles of the universe such as protons and neutrons). There is another class called leptons; they are the light particles of the universe. The electron is in this class, but more importantly, so is a particle called the neutrino. We will see later that the neutrino is now playing a central role in the dark matter mystery. Aside from this, though, there is another way around the nucleosynthesis argument. The calculations that lead to the predicted amount of deuterium are based on the assumption that there were very few neutrinos around at the time, and also that during nucleosynthesis the universe was isotropic (the same in all directions). Several astronomers, including Craig Hogan of the University of Arizona, James Applegate of Columbia University, Phillip Scherrer of the University of Chicago, and Martin Rees of England, have come to the conclusion that the universe may have been inhomogeneous during nucleosynthesis. If so, the predicted amount of deuterium (and other light elements) is incorrect.

In a recent paper Applegate and Hogan wrote, "Inhomogencities which persisted until the epoch of nucleosynthesis can lead to nuclear products with very different composition from the standard hot big bang. . . . These events may produce observable distortions in cosmic light-element abundances; in particular, the deuterium abundance may be increased by orders of magnitude."

But there are other light elements such as lithium, and they also seem to indicate the universe is open. I asked David Schramm of the University of Chicago if he thought it was possible to get around the baryon limit via inhomogeneity. "The lithium constraints now make such scenarios very improbable," he said. "The observed lithium–hydrogen ratio, which is approximately 10^{-10}, is only produced for a very narrow parameter range. Any mixing gives larger values and this disagrees with observation." He later told me that helium measurements are also now a problem.

Virginia Trimble of the University of California, who has also done considerable work on the dark matter problem, says, "The chief difficulty is clearly the lithium abundance. But it is possible it can be overcome." In summary, then, we can say that unless we can find a large amount of matter that is not composed of baryons, or else nucleosynthesis was "lumpy" enough to produce more deuterium, the universe is open by a large margin. And this, of course, means that it will expand forever.

THE FLATNESS PROBLEM

There is another argument that seems to indicate the universe is open. It was pointed out by Robert Dicke and Jim Peebles of Princeton University in 1979. They showed that any deviation from exact flatness should increase linearly in time. This meant that if there was even an extremely small curvature in the early universe, say at nucleosynthesis, the universe would now be extremely curved. But we know it isn't. Measurements tell us its present density is very close to 1 (it is between 0.1 and 2), and therefore, at most, it is only slightly curved. If we extrapolate these numbers back to nucleosynthesis, however, they give 1 to 15 decimal places (i.e., 1.000000000000000). In other words, the universe was *exactly* flat then. This seems to suggest that it is flat now and always was.

But why should it be flat? A possible answer came in 1980 from Alan Guth, now of MIT. His idea was based, in turn, on a theory that had been put forward in 1972 by Howard Georgi and Sheldon Glashow—a unified field theory that attempted to unify three of the basic fields of nature (the electromagnetic, the weak, and the strong nuclear fields). Guth was able to overcome several of the problems that had plagued the big bang theory by assuming the universe suddenly inflated in size about 10^{-36} second after the big bang. In other words, its expansion rate increased dramatically for a very short period of time. It grew about 28 orders of magnitude in 10^{-30} second. The theory, now called

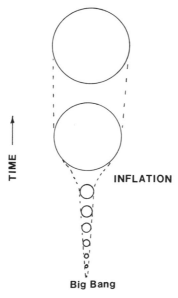

TIME →

INFLATION

Big Bang

Schematic of inflation.

"inflation theory," was later modified by A. Linde of the USSR and Paul Steinhardt and Andreas Albrecht of the University of Pennsylvania. One of the major predictions of the theory is that the universe is flat, which in turn means that its density is equal to the critical density. If true, there is a large amount of matter we are not seeing. Indeed, assuming the arguments from nucleosynthesis are valid, most of the mass of the universe must be in a strange form. It can't be neutrons or protons, or most of the other common heavy particles.

But is the inflation argument valid? I asked several people for their opinion. David Schramm replied, "Inflation provides a plausible mechanism for solving the flatness problem. Most likely the flatness argument is valid and something like inflation must have taken place, although the details may be quite different from the standard grand unified theory inflation. Any other density requires extraordinary fine-tuning just for our existence."

According to Virginia Trimble, "There are actually some forms of inflation that preserve many of its virtues and do not produce a flat universe. I am inclined to believe that the actual average density will eventually be determined by observation." And Peebles: "Inflation tells us the universe should be flat but it doesn't rule out the possibility that it is flat because there is a cosmological constant. That's pretty far out, I admit, but I'm beginning to think it's something we ought to bear in mind as a possibility."

C. S. Frenck of the University of Durham in England says, "It has at least a fifty percent chance of being correct."

And finally Ostriker: "I'm not at all convinced that the universe is flat. I am an agnostic on this 'religious' issue." But he admits his background in inflation theory is not strong. Like Peebles he also believes that if the universe is flat it could be caused by a cosmological constant rather than inflation.

OTHER PROBLEMS

Ostriker also reminded me that there are really three missing mass problems. Besides the one associated with galaxies, and the one associated with the universe, there's a problem associated with the prediction of nucleosynthesis.

"The third missing mass problem doesn't get much publicity," said Ostriker. "In essence, it's the fact that we can't account for all the baryons. If you ask what deuterium, helium, and lithium tell us about the average density of the universe, the answer is that it should be 0.1 times the critical density. But if you add up all the starlight you don't get 0.1. It only gives 0.01, so the stars don't account for all the baryons. Where are all the baryons that we know are there from nucleosynthesis? The fact that the number is so close to the dynamical number [density obtained from the dynamical motion of galaxies] makes me think that a reasonable possibility is that most of the dark matter is basically ordinary matter in some form that we haven't kept track of . . . low-mass stars, or black holes. But matter that is ordinary baryons." He paused briefly. "Of course I'm talking about the matter that's associated with galaxies," he said, inferring he wasn't referring to the missing mass problem of the universe. This led me to ask him if he thought the two problems were distinct. He replied, "In one case you know the matter is there but you can't identify it. In the other case you don't know it's there . . . in fact, you measure and you find it's not there. But you want to believe it's there. I'm convinced the problems are distinct."

Peebles answered the same question in the following way: "It's tempting to think they are related. If you had two missing quantities of mass that weren't the same. . . you'd think they might be related. But I would stress that they ought to be distinguished. One won't necessarily solve the other."

CHAPTER 6

Will the Candidates Please Stand

If from 90 to 99 percent of the universe is concealed from us we can't help but wonder: What form does it take? In the case of galaxies and clusters we know it is there. But why is it hidden from us? There are many possibilities, and many forms it could take. In this chapter I will examine all of them briefly. In later chapters I will look at some of the more important ones in more detail.

First of all, as I mentioned earlier, it is important to remember that we have two distinct problems: one associated with galaxies and clusters and one associated with the overall universe. It is perhaps best to refer to the first as the dark matter problem and the second as the missing mass problem, and I will try to adhere to this.

Strangely, even though we seem to be far short of the amount of matter needed to close the universe, only about one atom per cubic meter is actually needed. And when you think about it, this isn't much. If matter were uniformly spread throughout the universe with this density we would have considerable trouble detecting it. Distant objects would barely be dimmed. And atoms would rarely collide, so there would be little if any emission from them.

But would we expect such a gas throughout the universe? To answer this we have to go back to the early universe. We know that at one time all the matter of the universe was in the form of a huge gas cloud. This cloud eventually broke up into

smaller clouds, and within these smaller clouds condensation
into stars occurred. But did all the gas go into the making of
stars? It seems unlikely, for the star-making process is not that
efficient. It's quite possible that a considerable amount of material was left over.

Let's begin our survey of the candidates, then, by looking
at the dark matter problem in galaxies. A lot of what I say will
also apply to the missing mass problem but I'll leave the detailed discussion of that until the end of the chapter. One of the
major questions we have to ask ourselves in considering candidates for this case is: will it tend to "clump" around galaxies and
clusters?

THE DWARFS: RED, BROWN, BLACK, AND WHITE

Red dwarfs, as their name implies, are small stars, ranging in mass from about one-half to one-tenth that of the sun. Large numbers have been observed, but are the numbers large enough to account for the dark matter? That depends on their life cycle. We saw earlier that stars go through a cycle; large stars live only a few tens of millions of years, medium-sized stars like our sun perhaps 10 billion years, and tiny red dwarfs 50 or more billion years. In our discussion of elliptical galaxies I mentioned that massive stars explode as supernovae, with the resulting debris going into the making of new but generally less massive stars. This means that, in time, the average star in the universe will be relatively small—eventually, in fact, most will be red dwarfs. Indeed, if we look at the stars around us now we find that, by mass, about 80 percent of them are red dwarfs. Thus, red dwarfs are common; furthermore, they are exceedingly dim—only about one ten-thousandth as bright as our sun on the average—and they are therefore good candidates. But there are problems. Suppose, for example, that the red dwarfs were uniformly distributed throughout the galaxies in clusters so that clusters were ten times more massive than we now assume them to be. Collisions would then be more common, and within a few billion years all clusters would be significantly rearranged. The lowest mass galaxies, which would have the lowest speeds, would generally be in the outer parts of the cluster and the most massive galaxies would have fallen to the center.

Do clusters have these characteristics? Observational evidence indicates that they don't. One way around this is to assume the red dwarfs are in a huge extended halo around the galaxies. Of course, we saw earlier that this is just what Peebles and Ostriker predicted.

Another problem with red dwarfs is associated with the enormous number it takes to account for the dark matter. If there were, indeed, the required number, there should also be many stars about the mass of our sun mixed in with them. Yet

Some of the smaller stars in this cluster are dwarfs. (Courtesy National Optical Astronomy Observatories.)

we see no evidence for solar mass stars in the halo. In fact, it is easy to show that it would only take a few solar-mass stars mixed in with them to eliminate them as good candidates (the ML ratio would be too low).

A second type of dwarf is the brown dwarf. They have masses from about one-tenth the sun's mass down to slightly more than Jupiter's mass. They are, in effect, gas clouds that didn't quite become stars. They heated as they condensed, but their central temperature never became high enough to trigger nuclear reactions. Most of them likely shone for a while, then died away, and now emit no light.

It is reasonable to assume that a large number of brown dwarfs exist. They are only slightly smaller than red dwarfs (which we know exist). Furthermore, it's hard to believe that all condensations produced stars; many would no doubt have been too small. Oddly enough, though, we have not yet positively identified a single brown dwarf.

Another problem with both brown and red dwarfs, assuming they are in the halo, is that they would tend to make galaxies "sticky." This means that when they collide they would tend to stick together. And again we see no evidence of this. Furthermore, studies of "sky glow" indicate that brown dwarfs can't account for more than about 10 percent of the dark matter. So, although they have potential, there are many problems. We will see later, though, that they are the first choice of a large number of astronomers.

This brings us to white dwarfs. They arise when stars of medium mass like our sun die. All stars are in equilibrium under the influence of two equal and opposite forces: an inward one from gravity, and an outward one resulting from the gas pressure and radiation due to the thermonuclear furnace at the center of the star. When the thermonuclear furnace goes out the star collapses. In a medium-sized star the collapse is slow, taking millions of years. The end result is a white hot dwarf, only slightly larger than Earth. Several hundred such dwarfs have been observed, the best known being the companion of the bright star, Sirius. But compared to the number we would need, the number observed is low. And there is a reason for this: they are the remnants of stars at least as large as our sun, and it's unlikely that in the last 10–15 billion years a sizable number of such stars have died. On the basis of this we have to say that white dwarfs are not good candidates.

White dwarfs of course eventually cool and become black dwarfs. And they too are poor candidates. The universe is just not old enough for many—if any—to have formed.

Along the same line we have asteroids, comets, rocks, and dust. Perhaps the best of these is dust, but if there was enough to account for the dark matter it would be quite noticeable. It would, in fact, make space smoggy, and we see little evidence that it is.

HYDROGEN

This brings us to hydrogen. It has to be a good candidate; after all, 90 percent of the universe is made up of it. Since it occurs in three forms (neutral, molecular, and ionized) we'll consider each separately.

Let's begin with neutral hydrogen. Is there any way we can determine how much there is in the universe? Indeed, there is. Hydrogen absorbs at 21 cm and the spectrum of a galaxy would therefore have an absorption line at 21 cm. But the spectral lines of galaxies are also redshifted—the farther out the galaxy, the greater the redshift. This means that the light from a distant galaxy (or quasar) will absorb at many different wavelengths in its journey to us. And as a result the absorption line will be broadened into an absorption "trough."

Do we see such troughs? The light from both quasars and distant galaxies has been checked, and although some broadening occurs, it can be explained otherwise. On the basis of this it is reasonable to assume that there is little neutral hydrogen in space. Don't get me wrong though; there certainly is some, but not nearly enough to account for the dark matter.

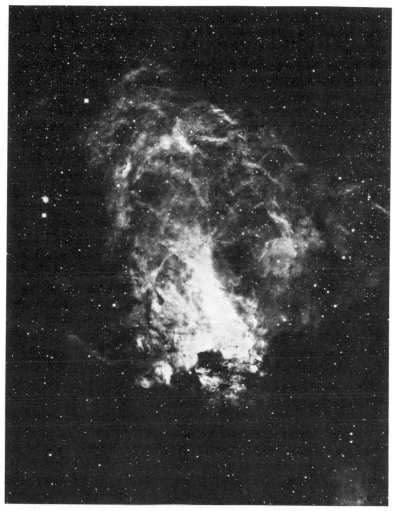

A gaseous nebula (made up mostly of hydrogen). (Courtesy National Optical Astronomy Observatories.)

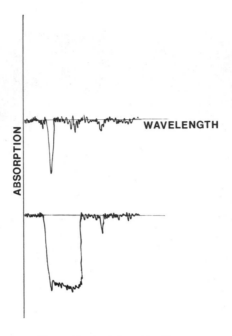

Upper: Graph of absorption line.
Lower: Graph of absorption trough.

What about molecular hydrogen (two hydrogen atoms bonded together)? The probability here is also low. Like neutral hydrogen it would also give rise to an absorption trough, and no troughs have been found.

The third form of hydrogen, namely hydrogen that has no electrons (ionized hydrogen), looks more promising, however. If hydrogen is ionized it must be at a high temperature—at least a few million degrees, and probably higher. And if the temperature is this high it would emit X rays. Do we see such X rays? We have indeed detected a large number of X rays sources in space, but of course many different things give off X rays. What is more important, though, is that astronomers have detected a cosmic background of X rays.

Astronomers first became interested in the possibility of X rays in space after World War II. Using a captured V-2 rocket scientists were able to show that the sun was a weak X ray source, but the X rays were so weak that no one thought we would be able to detect them from distant sources. Then in the early 1960s things changed.

Riccardo Giacconi, working for a research company called American Science and Engineering (which had been established a few years earlier to help monitor nuclear fallout from the atmosphere), decided to take another look.

Two problems confronted him: he had to design and build a detector, and he had to get it above Earth's atmosphere. The obvious way to get it above the atmosphere was via rockets, and this meant that it had to be small. But Giacconi was up to the task, and on June 12, 1962 his tiny detector was placed aboard an Arobee rocket and launched into space. The major goal of the project was to check solar X rays reflected from the moon. None were found, but as the detector scanned the sky beyond the moon Giacconi and his colleagues were delighted: a strong source in the constellation Scorpius was discovered. The most important discovery, though, was that regardless of where the detector was pointed there were X rays. In other words, there was a cosmic background of X rays—an "X-ray glow" from the sky.

What could be causing this? To answer this we must look at how X rays are generated. One of the ways is through collisions: if we have a mixture of high-speed electrons and protons, the negatively charged electrons will be deflected as they pass the positively charged protons, and in the process they will give off X rays.

Is this the way the X rays from space are generated? It's possible if we have a mixture of electrons and protons out there—what we call a plasma. But the plasma has to be hot—extremely hot. Temperatures of the order of 500 million degrees are needed, and this is where the problem comes in. What kind of mechanism would generate such high temperatures? Astron-

omers are still not certain; furthermore, even if they were generated, why hasn't the gas cooled off? There seemed to be so
many problems that some astronomers wondered if the X-ray
background was really there.

But in 1970 another X-ray satellite, Uhuru, was launched.
Four hundred new X-ray sources were discovered—and the
background X rays were, indeed, verified. Another important
discovery was that there was intergalactic gas in and around
clusters of galaxies, and this gas emitted X rays. There was, indeed, plasma in space.

Further confirmation and even more discoveries came in
1978 when Giacconi and his colleagues launched the tremen-
dously successful X-ray satellite, Einstein. It provided the first
true X-ray images of the universe. But when the first photos
were released there was a shock: thousands of X-ray sources
were found. Regardless of where the detector was pointed
dozens of sources showed up in each square degree of the sky.
Maybe the cosmic background wasn't from a plasma throughout
the universe, after all. Astronomers are, indeed, now convinced
that most of the X rays are from quasars. A plasma might be
responsible for some but we're still not sure how much.

STELLAR CORPSES

We have already talked about one of the major stellar
corpses of the universe—the white dwarf. But there are others.
White dwarfs occur when stars about the mass of our sun die.
What about stars even more massive? If the star is over about
four solar masses it ends its life with a tremendous explosion—
the supernova. Most of the star is blown off into space, but
strangely, the explosion compresses a small amount of the material near the core. The particles are crushed until almost nothing
but neutrons remain—the result is a tiny neutron star only a few
miles across.

We have identified several hundred neutron stars, or
pulsars as they are now called, and there are no doubt thou-

sands we cannot see. So again we ask: are neutron stars good candidates for dark matter? After all, they are extremely small, and many are likely undetectable. Our answer has to be no, for the same reason that white dwarfs are not good candidates. In fact, the problem is even more serious in this case. It takes a massive star to produce a neutron star, and there are just not enough massive stars around. And it's unlikely there ever were. We have another stellar corpse that is of considerable interest. In fact, it is of so much interest that I will spend an entire chapter discussing it later so I will say little about it here. I'm talking about black holes: strange, exotic objects that result when supermassive stars (> ten solar masses) collapse. Many were also presumably created in the big bang explosion that created the universe. Most interest has, in fact, centered around this type, called primordial black holes. We will look at them in considerable detail in the next chapter.

THE RADIATIONS

It may surprise you that I'm using the plural in the title of this section. Most people are familiar with only one type of radiation, namely electromagnetic radiation. But there is another type: gravitational radiation.

Let's begin with electromagnetic radiation. This is the ultraviolet, infrared, visible, radio radiation and so on that you are likely familiar with. But why would we consider it a candidate for dark matter? After all, we're looking for mass. But, as Einstein told us many years ago, energy and mass are equivalent, and electromagnetic radiation (photons) is energy. This means that there is a mass-equivalent for all the electromagnetic energy in space. How large is it? Or putting it another way: how many photons are there in space? There must be a large number; after all, the universe is literally bathed in the photons left over from the big bang (the cosmic background radiation). Simple calculations show us that there are 10^8 photons for every baryon in the universe. This sounds like a lot, but unfortunately when we convert them into energy we find they don't contribute much—only one ten-thousandth the amount needed. Furthermore, if we're talking in terms of dark matter in galaxies and clusters we have the problem of how photons would clump. They don't—the cosmic background is uniform.

Now for gravitational waves. There is a serious problem here but it may one day be overcome. Even though such waves are predicted by general relativity they have never been detected.

There was a time, however, when we thought they had. Joseph Weber, working at the University of Maryland, announced in 1969 that he had obtained direct evidence of their existence. The announcement astounded the scientific world. But when other experiments were set up to check on his results they weren't verified. And to this day we still have not detected gravitational waves directly.

This is strange in that gravitational waves literally have to exist; there seems to be no way around them. To see why, consider the gravitational field of, say, Earth. If something suddenly crashed into Earth, its mass and therefore its gravitational field would change, and something would have to tell each point in space around it that a change had occurred so appropriate adjustments could be made. A similar situation exists in the case of electromagnetic waves. When a change in an electric charge occurs somewhere, an electromagnetic wave propagates out to effect a change in the electric field around it. It seems reasonable to assume that this would also happen in the case of gravitation.

Gravitational waves are, indeed, quite similar to electromagnetic waves. An electromagnetic wave is generated when a charge oscillates; in the same way a gravitational wave should be generated when a mass oscillates. But there is a major difference between the two: the electromagnetic wave is 10^{37} times stronger than the gravitational wave.

Does this mean that gravitational waves are too weak to be

detected? For most gravitational waves the answer is yes, since (in theory) a gravitational wave is emitted even when two small masses at the ends of a spring are oscillating. The waves that are given off in this case, however, are far too weak for detection. But there are cases where relatively intense gravitational waves would be generated. Supernova explosions might be an effective source. Better still would be a system of two extremely dense objects—say two neutron stars revolving around one another. Or perhaps a neutron star and a black hole.

And, indeed, such a source has been found. In 1974 J. Taylor and R. Hulse discovered a binary system on the border of the constellations Sagitta and Aquarius that appeared to be composed of dense objects. One of the two is believed to be a pulsar, the other perhaps a black hole. It is referred to as the "binary pulsar."

Although gravitational waves have not been detected from the binary pulsar directly, they have been detected indirectly. As the two objects revolve around one another they lose considerable energy, and this loss shows up in their orbit: it decreases in size. Analysis of the orbit has shown that in addition to electromagnetic energy a small amount is being given off in the form of gravitational waves.

But are gravitational waves a good candidate for dark matter? Because we know so little about them, and they haven't been verified directly, we have to say no.

EXOTIC PARTICLES

If the predictions of nucleosynthesis and inflation theory are correct most of the dark matter must be in a form other than baryons. Candidates in this category are usually referred to as exotic particles; they include neutrinos, axions, gravitinos, photinos, monopoles, and a number of other particles. I'll only talk about them briefly here because they will be discussed in detail in a later chapter.

Let's consider the neutrino first. For many years it wasn't considered to be a serious candidate, mainly because it was assumed to have zero rest mass. Besides, it was extremely difficult to detect. In the mid-1970s, it was suggested that the neutrino might have a small mass; if so it would be an excellent candidate because there are such a large number of them (they are as numerous as photons). Experiments were set up and in 1980 a Russian team announced they had measured the mass. Things looked bright. But when others tried to verify the result, they couldn't. Today, we're still waiting for the final word.

Next we have the monopole. The best way to describe a monopole is to consider a magnet. As you likely know, a magnet has a north and a south pole. If you cut it in half you get another north and south pole; in fact, if you continue cutting it ad infinitum you continue getting two poles. But do single north and south poles exist in nature? It seems reasonable that they do; after all, we have particles with single electric fields. Why not particles with single magnetic poles? Such a particle would be a monopole, and many scientists are convinced they do exist. Theory, in fact, predicts that they should, but so far no one has positively identified one. And until they do, they must be considered a shaky candidate.

Other possibilities exist among a number of particles predicted by supersymmetry, a theory that was devised in 1975 by Bruno Zumino who was then at CERN and Julius Wess of the University of Karlsruhe in Germany. According to this theory massive particles called gravitinos, photinos, winos and so on exist. The major problem is that none of them have been detected; they presumably lie just beyond the range of our present accelerators. If they are eventually found, though, they could be excellent candidates.

In the same category are axions, exceedingly light particles—a trillion times as light as the electron. They could easily settle around galaxies creating a halo. But again they exist only on paper.

A final candidate worth considering isn't a particle at all. It's a string—an extremely heavy string called a cosmic string. According to theories of the early universe these strings may have been created as the universe went through a series of phase changes. Just as steam condenses to water, then freezes to ice, there may have been similar phase changes in the early universe. These strings would be exceedingly massive—a piece an inch long would weigh as much as Earth, and some would stretch from one end of the universe to the other. Most, however, would be in the form of gigantic loops. But, according to Virginia Trimble, "They would likely contribute very little to the mass–energy density of the universe." And, of course, we're still not certain they exist.

Thus, we see that the major problem with all the exotic particles (with the exception of the neutrino) is that we're not sure they exist.

THE MIMICS

It is possible that there is no dark matter in galaxies and clusters, and that something else in the universe is making it appear there is. In other words, something is mimicking dark matter. One possibility is a changing gravitational constant; if this constant increases with time or varies suitably with the acceleration of the universe, the mass of galaxies and clusters will appear to be larger than it really is. And there are theories that do indeed predict that G (the gravitational constant) is varying. The best known is the Brans–Dicke theory, a theory put forward in 1961 as an alternate to Einstein's general theory of relativity.

So far, though, we have found little evidence for a varying G. And most of the tests that have been devised seem to rule out the Brans–Dicke and other similar theories.

A second possibility is related to a constant called the cosmological constant that Einstein introduced. He added it to his cosmological equations because he found that without it the universe was unstable. When it was later found that the universe was expanding Einstein discarded the constant. But others have kept it, and it is still seriously considered today in some theories.

MISSING MASS OF THE UNIVERSE

So far we've been talking mainly about dark matter in galaxies and clusters. Much of what I said, though, applies to the missing mass problem of the universe (assuming there is some missing mass). Whether the dark matter of galaxies constitutes some of this missing mass is still unknown. But even if it does, it seems unlikely that it is enough. If galaxies in clusters contained enough mass to close the universe, they would have, on the average, much higher speeds than they do. Furthermore, ML ratios would have to be 1000 or greater, and there is no evidence that they are.

It seems most likely that the missing mass would be spread uniformly throughout the universe. And some of the candidates we discussed earlier could, indeed, be distributed uniformly. Neutrinos are one possibility. But it turns out that if they have the required mass to close the universe their speeds would be so slow they would likely cluster around galaxies—perhaps fall into their centers. An even better candidate in this respect is ionized hydrogen. We know that it emits X rays, and an X-ray background has been found. Much of it, though, no doubt comes from quasars. So again, we're uncertain.

Another possibility is stars between the galaxies. There are likely some there—ripped off when galaxies collided. Furthermore, it is quite possible a few strays were formed there. But it is extremely unlikely that 99 percent of all stars are outside galaxies. So they are not a good candidate. Most of the other things we considered earlier—black holes, dwarf stars, gravitational waves, and so on—also have to be ruled out as candidates for the same reason they do for galaxies.

OPINIONS

In an effort to find out what some of the favorite candidates were I asked several prominent astronomers. Their replies were as follows.

David Schramm: "The tau neutrino. . . . Neutrinos still do the best job on large-scale problems."

Jim Peebles: "If you're talking about the matter within galaxies my favorite is low-mass stars." I then asked him about exotic particles. "I've discovered that as I grow into the golden years I'm getting more and more conservative. So among the exotic particles I would have to put most of my money on massive neutrinos. Neutrinos do exist."

Jeremiah Ostriker: ". . .Basically ordinary matter [matter made up of protons and neutrons] . . . in the form of low-mass stars or black holes."

Michael Turner of the University of Chicago: "My top three picks are axions, massive neutrinos, and photinos. They are the three that show the most promise." I asked him about monopoles. "My dark horse candidate would have to be the monopole. I have a soft spot in my heart for them."

Vera Rubin: "Plain ordinary matter . . . very low-mass stars or something rather 'ordinary' would please me. But nature is not here to please us, and as an observer I am prepared to accept whatever is found. If not ordinary matter, it would be fun to have it turn out to be something we cannot yet even speculate about. Open-mindedness is a necessity."

Craig Hogan of Steward Observatory in Tucson: "Stellar remnants, especially black holes. It seems a very natural thing to have large amounts of dark matter left over from early generations of stars. Another good candidate would be failed stars or 'Jupiters.'"

Virginia Trimble: "How am I betting? At least fifty-fifty on protons and neutrons all the way. The remaining probability is distributed very thinly among other exotic dark candidates."

Geoffrey Burbidge: "Probably very low-mass stars or planetary bodies."

Jan Oort: "Quite likely candidates . . . are low-mass stars."

So there we have it: an overview of most of the candidates. In the next few chapters I will take a more detailed look at some of the more seriously considered ones.

Summary of Candidates

Candidate	Positive features	Problems
Red and brown dwarfs	Red dwarfs are common Brown dwarfs are expected to be common	Large numbers are required Brown dwarfs have not been detected
White dwarfs	Have large masses for their size	Not common enough
Neutron stars	Large mass There are likely many we do not see	Created from massive stars, which are rare
Hydrogen (neutral, molecular, ionized)	Most common element in the universe	21-cm trough not seen
Black holes	Many primordial black holes were likely created in the big bang	Stellar collapse black holes rare
Massive neutrinos	Predicted by grand unified theories (GUTs) Can clump around galaxies	Mass not verified
Magnetic monopoles	Very massive Predicted by GUTs	Have not been detected
Supersymmetric particles	Predicted by supergravity	Have not been detected
Axion	Can clump around galaxies Predicted by GUTs	Has not been detected

CHAPTER 7

Black Holes

We have seen that black holes are a serious candidate for the dark matter. But I have said little about what they are. In this chapter we will look at their properties, and consider them as candidates.

THE DISCOVERY

On November 27, 1783, Reverend John Michell presented a paper to the Royal Society of London suggesting that giant "invisible stars" might exist. The announcement caused considerable excitement. Was it possible? Some thought it was just idle speculation. But Michell had made detailed calculations using Newton's law of gravity. And according to these calculations a star 500 times as large as the sun had a gravitational field so strong that light could not leave it. All we would see is a giant black sphere, visible only because it blocked the light from stars behind it.

A similar suggestion was made by Pierre Laplace of France a few years later in his book *The System of the World*. His calculations, however, led him to a star somewhat smaller than Michell's—about 250 times the size of the sun. Laplace apparently had second thoughts about it, however. After the wave theory of light became popular he deleted it from his book.

Both men had the right idea, but both used Newton's law of gravity, and both were thinking of a large star. So, although

What a black hole might look like if you encountered one in space.

their object has some of the properties of a black hole, it is different in most respects. The black hole of today is not a gigantic star, but the remnant of a gigantic collapsed star. Furthermore, it is described by Einstein's theory of general relativity.

Because general relativity plays such an important role in black hole physics let's take a few moments to look at it. Einstein published the theory in 1916 after a struggle of several years. It was the extension of an earlier theory he had published in 1905—the special theory of relativity. General relativity is, in essence, a theory of gravity. It is a complicated theory, based on complicated mathematics; in fact, the equations were so complicated that even Einstein couldn't solve them. The first solution came from the German astronomer Karl Schwarzschild. And it came, not at a desk at a university as you might expect, but in a bed on the Russian front during World War I. Schwarzschild was strongly patriotic, and although as a scientist he was exempt from the army, he had volunteered and was sent

Karl Schwarzschild.

to the Russian front. While there he contracted a rare disease that confined him to bed. Crippled, and in a weakened state, he received a copy of Einstein's theory, which he read with interest. Noticing that Einstein had only obtained an approximate solution he began looking for an exact solution. And he found one. He quickly sent it to Einstein, who was pleased upon receiving it. Einstein wrote back, "I had not expected the exact solution of the problem would be formulated so simply. . . ." Within days he presented it to the German Academy.

Schwarzschild managed to get two papers published through Einstein, but he didn't live to see their success and the fame they eventually brought him. For shortly after communicating with Einstein he died.

In the second of his papers Schwarzschild pointed out that in certain cases the solution was particularly strange. Einstein was also disturbed by this. Let's take a moment to look at the details. I'll begin by defining what is called a singularity. It is a place where the solution (in the above case, the solution of Einstein's equations) becomes infinite. This frequently occurs at a radius of zero. For example, in Newton's theory of gravity the gravitational field of a point particle becomes infinite at a radius of zero. Similarly, the electrical field around a charge is infinite at zero radius. So it was perhaps expected that an infinity would appear in Einstein's equations at zero radius. But what surprised them was that there also appeared to be an infinity—a singularity—at a small finite radius. Schwarzschild calculated this radius to be 3 km for the sun. But he also showed that a static sphere of fluid of uniform density would reach infinite pressure upon compression before it reached this radius—called the gravitational radius. He was therefore not disturbed by it. Today, of course, we realize that this singularity predicts the existence of a black hole. But early scientists were confused over its meaning.

Shortly after Schwarzschild obtained his solution a Dutch scientist, Johannes Droste, obtained the same solution. He also noticed the singularity at the gravitational radius, and he went further than Schwarzschild in calculating its properties. He determined the orbits of particles and light rays around it, finding that just outside it light rays would take up circular orbits. He also determined that, as seen by an outside observer, a point particle would take an infinite amount of time to fall to this surface. Still, he had little idea what it meant.

For years scientists grappled with the physical significance of the gravitational radius, and the surface associated with it. Was it real? Could it be realized in nature? Or was it just a mathematical anomaly? The German mathematician Hermann Weyl looked carefully at the geometry of the surface and the space surrounding it, and came surprisingly close to a correct answer. Still, the details eluded him. And even though others also came close, it wasn't until 1933 that the first published statement came that the singularity at the gravitational radius was only a "fictitious" singularity. But the announcement was buried in a cosmology paper by Georges Lemaître and few saw it. One who did was H. P. Robertson, an American cosmologist. He brought it to the attention of Einstein.

This led Einstein and a colleague, N. Rosen, to reexamine the geometry around a dense object. According to general relativity the space around a massive object is curved. They found that the curvature increased as you got closer to the object. It was like a tunnel of decreasing diameter: it got smaller and smaller as you approached the black hole. To their surprise, though, they found that there was a mirror image tunnel on the other end of the black hole. In effect, the tunnel passed beyond the black hole. Einstein wondered where you would end up if you ventured into this tunnel, or Einstein–Rosen bridge as it was later called. The only answer he could come up with was: another universe. And he didn't like it. He was therefore pleased when, upon performing the appropriate calculations, he found that it took a speed greater than that of light to get through, and such speeds were impossible according to special relativity.

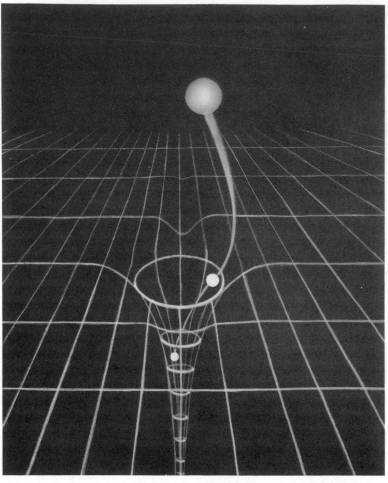

The "throat" of a black hole.

Then came 1939. Robert Oppenheimer of the University of California decided to use Einstein's general theory of relativity to look at the collapse of a star. We saw earlier that a star is born, lives for a few billion (or million) years depending on its mass, then dies. Its death is signaled by the last flicker of its thermo-nuclear furnace, and a collapse as gravity overcomes it. Oppenheimer was determined to find out what happened to the star after the collapse.

A few years earlier (1933) Fritz Zwicky and Walter Baade of Mt. Wilson Observatory had suggested that an exotic star composed mostly of neutrons—a neutron star—might be left after a supernova explosion. Such objects had been talked about almost from the day that Chadwick discovered the neutron in 1932. Shortly after the discovery, Lev Landau of Russia, while talking to Niels Bohr and Leon Rosenfeld, mentioned that such a star might exist. He later wrote a paper outlining some of its properties. George Gamow also became intrigued with the idea, discussing it at length in his scientific and popular books.

It is perhaps strange that Oppenheimer embarked on such a problem. General relativity was, for the most part, outside his mainstream of interest. In 1939 he had just returned from Europe where he had been caught up in the latest discoveries in quantum theory. When he came to Berkeley he was one of the few Americans who knew anything about the new theory. As a result many students flocked to work under him. Most of his work at this time was on quantum theory and nuclear physics, but when George Volkoff, who had earlier emigrated to Canada from Russia, came to work under him in 1938 he gave him a problem in general relativity. He suggested that Volkoff use the theory to see if stable neutron stars existed. Volkoff found that there were, indeed, stable neutron star configurations with masses that ranged between one-third and three-fourths of a solar mass (these limits were later adjusted). On February 15, 1939, his results were published in a paper titled "On Massive Neutron Cores."

This was a significant result. But an even more astounding result came a few months later when Oppenheimer and another student, Hartland Snyder, published a paper that predicted the existence of an "ever-collapsing star"—what we now call a black hole. Titled "On Continued Gravitational Contraction," this paper would become one of the most important ever published in black hole physics. The abstract begins with the statement, "When all thermonuclear sources of energy are exhausted a sufficiently heavy star will collapse. Unless forces due to rotation [stopped it] . . . this contraction will continue indefinitely . . . the radius of the star approaching . . . its gravitational radius."

A startling statement. Indeed, it was: it meant that a star, if sufficiently massive, would collapse forever. How was this possible? It was certainly difficult at that time (and even now) to understand. It was, in fact, so astounding that Oppenheimer didn't quite know what to make of it. Writing to another scientist he said, "We have been working on static and nonstatic solutions for very heavy masses that have exhausted their energy sources; old stars which collapse . . . . The results have been very odd. . . ."

Although the 1939 paper eventually became a classic, it left many questions unanswered. What happened to the mass of the star as it collapsed inside its gravitational radius? What went on inside this mysterious surface?

Neither Oppenheimer, Snyder, nor Volkoff pursued the problem. World War II broke out within months and soon all were working on other problems. Oppenheimer became involved with the Manhattan Project. Snyder went to Northwestern University, and Volkoff, after a brief stay at Princeton, went to the University of British Columbia in Canada.

The objects predicted by Oppenheimer and his students were exotic—so exotic that few took them seriously. It seemed impossible that they would actually exist. Neutron stars were bad enough—but black holes were crazy. Only Zwicky of Mt. Wilson seemed interested; he continued to publish papers on neutron stars. But nobody cared.

Even when the war ended, no one took an interest. Not until several years later did John Wheeler of Princeton begin to wonder about them. He dug up the papers of Oppenheimer,

Snyder, and Volkoff. But he was not convinced they were correct.

Wheeler obtained his Ph.D. at Johns Hopkins in 1933. From there he went to Europe to learn quantum theory, finally returning to Princeton where he remained until he retired. He made several important contributions to atomic and nuclear physics but his first love was general relativity. The Oppenheimer–Snyder paper impressed him, but he wasn't convinced that an ever-collapsing star was possible—the entire idea seemed to be against the laws of nature. Incidentally, it was John Wheeler who invented the name "black hole" for the object.

In 1957 Wheeler and two of his students, Kent Harrison and M. Wakano, began writing a computer program that would determine the end result for the collapse of a star of any mass. Using the huge MANIAC computer at Princeton, they ran the program and—lo and behold—they got not only the Oppenheimer–Snyder results for the black hole, but also the

John Wheeler.

Oppenheimer–Volkoff results for neutron stars, and even corresponding results for white dwarfs—all in one swoop.

Wheeler presented the results in 1958 at a conference in Brussels. But few were interested. One who realized the significance of the results, however, was Yakov Zel'dovich of the Soviet Union. He was already becoming a respected scientist in the area and would soon build a powerful Soviet school of astrophysics and general relativity. The Soviets had not tucked the Oppenheimer papers away in their closet the way Americans had; they were accepted, well-known results. And because of their interest they soon had a commanding lead over Americans in the area. Except for Wheeler no one in the United States was interested.

But an important development did occur in America about this time. Scientists had, for years, been struggling with the singularity associated with the gravitational radius, realizing finally that it was just a "fictitious" singularity. Using an appropriate coordinate system it could be transformed away. The proper system was discovered by the Princeton physicist Martin Kruskal. In the mid-1950s Kruskal and a number of colleagues set up a study group to learn general relativity. Kruskal became intrigued with the singularity at the gravitational radius. He soon found he could make it disappear using a transformation to a new coordinate system. He showed it to Wheeler, but Wheeler seemed to take little interest—sure it was of no importance. So Kruskal never bothered to publish it. But by word of mouth it became known to others who discussed it in their papers. Finally Wheeler realized it was an important advance, but Kruskal still didn't bother to publish it. Wheeler therefore took the bull by the horns and published it himself— giving credit to Kruskal. About this time it was discovered independently by George Szerkeres of Australia.

Stellar collapse and black hole physics were still in the doldrums in America. Then came a discovery that shook up both astronomy and physics: strong radio sources, called quasars, were detected. What were they? They seemed to have an energy output beyond imagination. Astronomy was in a state of turmoil. An explanation had to be found. Where was all the energy coming from? Gigantic collapsing stars—superstars—seemed to be one of the best bets. Fred Hoyle and William Fowler were the first to make the suggestion. It came in the March issue of *Nature*—interestingly, in the same issue that Martin Schmidt and Jesse Greenstein of Mt. Wilson showed that quasars were in deep space, beyond the outermost galaxies.

By June it had become obvious that a meeting was needed to discuss the new discoveries. It was called by Peter Bergmann, Ivor Robinson, Alfred Schild, and E. L. Schucking. Invitations were sent to 300 scientists for the *First Texas Symposium on Relativistic Astrophysics* in Dallas, December 16–18, 1963.

All the notables of astronomy and relativity were there: Fowler, Hoyle, the Burbidges, Sandage, Wheeler, Chiu, and many others. And literally everyone was caught up in the excitement of the new discoveries. There was a feeling that it was a historic meeting—and indeed it was.

Hoyle and Fowler talked about their gravitational collapse models of quasars. Jesse Greenstein and Martin Schmidt presented the latest information on quasars. Hong-ya Chiu talked about neutron stars. Burbidge and Sandage presented a paper on violent events in the nuclei of galaxies. And, of course, there was a mammoth paper by Wheeler and his students. It was so long that it was not published with the regular proceedings, but rather as a separate volume.

The highlight of the meeting for many of the relativists was a paper by Roy Kerr of the University of Texas. It was one of the shortest papers, but also one of the most important. For years scientists, Einstein among them, had been struggling to solve the general relativistic field equations for the case of rotating matter (e.g., a rotating star). But no one had succeeded—until now. Kerr presented a solution showing that the surface of the

Roy Kerr.

black hole arising from the collapse of a spinning star was much more complicated than the nonspinning case. Like the nonspinning variety it also had a spherical surface at the gravitational radius; this surface is called the event horizon. But in a spinning, or Kerr, black hole there is another surface outside the event horizon called the static limit. These two surfaces touch along their axis of rotation, and between them is a region called the ergosphere. We'll talk about it in more detail later.

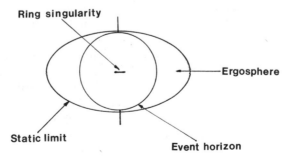

Internal structure of a Kerr black hole.

The Texas meeting provided the impetus the Americans needed. After it was over many relativists began to work in the area of black hole physics, and the field blossomed quickly. But the Soviets, under Zel'dovich, still had a commanding lead.

Theorists were finally starting to get excited about the possibilities. But there was still the question: did such objects—black holes—actually exist? No evidence for them had been found. In fact, the less exotic objects—neutron stars—had not yet even been found.

Then in 1968 came the announcement that strange pulsing objects had been discovered. While looking for quasars Antony Hewish and Jocelyn Bell of Cambridge University found what would later be called pulsars. Scientists soon showed that they had to be small and dense. But no one was sure what they were. White dwarfs seemed to be the best candidate. But then a pulsar was found in the Crab Nebula that rotated at 30 times a second. Too fast for a white dwarf. Tommy Gold of Cornell had been convinced from the beginning that the objects were neutron stars, rather than white dwarfs. And with the discovery of the Crab pulsar he was able to prove it.

Neutron stars existed. The odds on the existence of black holes immediately shot up. But how would we detect them? Certainly we wouldn't be able to see them directly; they were far too small—only a couple of miles across. A black hole in a cloud of gas might be visible, since it would pull the gas in and emit radiation in the process. But easier to identify would be one in a binary system. If one of the two stars collapsed to a black hole it might, in time, begin to pull material from the other star into it. If so, X rays would be generated.

Interestingly, a suspicious X ray source had already been discovered in the constellation Cygnus in 1964. But all attempts to locate the source optically had failed. Then in 1971 the X ray satellite Uhuru was launched and it soon zeroed in on the Cygnus source, now called Cyg X-1. The location was pinned down accurately and astronomers began their search. Soon a

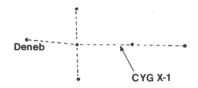

Position of Cyg X-1 in the constellation Cygnus.

star called HD 226868 was identified as close to the source. But was it the source itself?

Paul Murdin and Louise Webster of the Royal Greenwich Observatory obtained a spectrograph of the star within days. It was a blue supergiant at a distance of 6000 light-years that had a mass 23 times that of our sun. But were the X rays coming from it? It seemed unlikely; blue supergiants are not usually X-ray sources.

Webster and Murdin took more spectrographs. Soon it was evident that the lines were shifting back and forth: HD 226868 had to have an invisible companion. With a knowledge of the mass of HD 226868 they were able to estimate the mass of the invisible companion. It appeared to be at least ten solar masses—easily massive enough to be a black hole (only three solar masses is required). Study of the X rays revealed that their source could be no larger than about 20 miles across. It appeared as if the collapsed object—possibly a black hole—was attracting material from the outer regions of HD 226868 and emitting X rays as it spiraled in. And although several alternative models have been put forward by skeptics over the years we are now about 95 percent certain that Cyg X-1 does, indeed, contain a black hole.

While observational astronomers struggled with Cyg X-1, theoreticians continued to make important advances. A new star was, in fact, emerging: Stephen Hawking. Although struck down with ALS (a crippling nerve disease) and confined to a wheelchair while a graduate student at Cambridge, Hawking was still able to make important contributions. Working with

Roger Penrose of Oxford he showed that once an event horizon formed a singularity inevitably formed within it. Werner Israel, a Canadian physicist with a doctorate from Ireland, picked up on the discovery and showed that nonrotating black holes had to be perfectly spherical, and their size depended on their mass. Hawking and Brandon Carter of Cambridge University generalized this to include Kerr black holes, and later it was found to include all types of black holes. John Wheeler whimsically referred to it as the "black holes have no hair" theorem. What it says, in effect, is that black holes can have only three properties: mass, spin and charge. Anything else is lost by the star when it becomes a black hole.

PRIMORDIAL AND EVAPORATING BLACK HOLES

Even though many of the advances were being made in Europe, John Wheeler, in America, was still hard at work. Students were flocking to work under him; among them was Demitrios Christodoulou. Roger Penrose in England had just shown that energy could be extracted from a black hole if a particle was ejected into its ergosphere. (If the particle split with one part going into the event horizon, and the other emerging, the emerging one would carry off considerable energy.) Wheeler encouraged Christodoulou to study this process and find out what he could about it. Christodoulou soon discovered that energy could not be extracted indefinitely. There was, in effect, a state from which no further energy could be extracted; he referred to it as the "irreducible mass."

In 1970 Hawking included Christodoulou's result in a theorem showing that the surface area of a black hole could never decrease. Then things took a surprising turn. With Hawking's theorem, and the "no hair" theorem, black hole physics was beginning to look surprisingly like an older well-established branch of physics called thermodynamics. Another of Wheeler's students, Jacob Bekenstein, decided to push the analogy by advocating that, aside from a constant, the area of a black hole was the same as an important concept in thermodynamics called entropy. (Entropy is a measure of the disorder of a system.) Bekenstein's idea was radical—so radical that few took it seriously. "I was often told I was headed in the wrong way," he wrote, "but I took some comfort from Wheeler's opinion that black hole thermodynamics is crazy, perhaps crazy enough to work."

The major difficulty with the idea was that it meant that the black hole had to have a surface temperature greater than 0 K. To almost everyone this was blasphemy. Black holes were perfectly black; they absorbed everything, and emitted nothing, and therefore couldn't have a temperature above zero. Bekenstein, himself, was unsure of the significance of the temperature, referring to it as unreal.

One of the most outraged by the suggestion was Hawking. He acknowledged that there was an analogy between the two sciences, but as far as he was concerned that's as far as it went. Any talk of a real surface temperature, or a real entropy, was nonsense. R. Geroch of the University of Chicago, in fact, had demonstrated through a simple "thought" experiment that the entropy could not be real.

In 1973, however, Hawking visited the Soviet Union and talked to Yakov Zel'dovich and Alexander Starobinsky about black holes. They had begun to apply quantum theory to them, and had made some important discoveries. "They convinced me that . . . rotating black holes should create and emit particles," he said. Upon his return to England he began developing the details and discovered that any type of black hole would emit radiation and particles. This disturbed him; his first thoughts were that Bekenstein would use it to prove his case for black hole entropy. But the more Hawking looked at the calculations, the more he became uneasy about his stance against Bekenstein. "Maybe he was right, after all," he thought. Then he calculated the spectrum of the radiation—and was shocked! It was exactly the same formula Planck had derived many years earlier for a perfect absorber. This was, in effect, the formula that had launched quantum theory.

In 1974 Hawking announced his result: black holes would emit particles and radiation, and may even eventually explode. "Rubbish" was the universal response. It was impossible. But as scientists looked closer at his derivation some of them were convinced. Other papers appeared reinforcing his results, and soon there was no doubt—he was right. Black holes could emit radiation, and they did have a surface temperature above zero.

But for stellar-collapse black holes (those arising from the collapse of a star) the temperature was barely above zero. It was so small it was insignificant. A couple of years earlier, though, Hawking had pointed out that a different type of black hole might exist—created in the big bang explosion that created the universe. If this explosion was inhomogeneous—and it had to have been (otherwise galaxies wouldn't exist)—black holes of all masses should have been created. They are now called primordial black holes. Some may have been as small as an atom, others gigantic. Calculations showed that the tiny ones, sometimes called miniblack holes, could have a high surface temperature. In fact, the smaller they were, the higher their surface temperature, and the greater the amount of radiation they emitted.

Hawking pointed out that the radiation was a result of "pair production" just outside the event horizon, a process in which particles of opposite charge are created out of the vacuum. If one of the two particles falls into the black hole and the other escapes, the black hole will emit particles and radiation.

Of course, if it radiates it loses energy, or equivalently, mass. This means it gets less massive, and as it does it radiates even more. And as it gets smaller it radiates more and more, until finally in the last seconds of its life it explodes. All black holes, in fact, eventually explode. Large stellar-mass ones, however, last hundreds of billions of years, but tiny ones— miniblack holes—last only seconds. One with a mass of 10^{10} g, for example, will last only three seconds. One with a mass of

10^{15} g, on the other hand, will last ten billion years. But ten billion years is roughly the age of the universe. They should therefore be exploding right now. Is there any evidence that they are? So far we haven't found any. We would easily be able to detect one if it exploded nearby—for example, within the solar system. But we haven't.

BASIC PROPERTIES AND CANDIDATES

A black hole would be a strange sight if you encountered one in space. All you would see would be a black circular region, visible only because it blocks the light of background stars. In

Famous last words: "I'd like to get just a little closer for a better look."

the case of a stellar-collapse black hole it would be a few miles across. By the time you were able to see it, though, you would already be under its powerful gravitational pull. And if you ventured too close you would be pulled in through its event horizon. Once through this surface there is no escape; you would need a velocity greater than that of light to get out, and relativity tells us this is impossible.

At the center of the black hole is the singularity—a region of infinite density. This is the remnant of the collapsed star. In a nonspinning black hole it is a point, but in a Kerr black hole it is a ring. As I mentioned earlier the Kerr black hole also has an additional surface outside it—the static limit.

If you did, indeed, fall toward a black hole, you would notice many strange things. The passage of time, for example, is affected in the neighborhood of one. The best way to explain this is to consider two observers A and B at some distance from a black hole. Assume that B falls toward the black hole. Observer A will notice that B's watch runs slow; in fact, it will almost stop as he nears the black hole's surface. Observer B will also encounter strong "tidal forces" that will tend to stretch his body as he falls. They are caused because the gravitational field is much greater on his feet than on his head (assuming he is falling feet first).

Something else you might be wondering about: How many different types of black holes are there? According to the "no hair" theorem, the only properties a black hole can have are mass, charge, and spin. This gives four types: ordinary or Schwarzschild (no charge or spin), Kerr (spinning), Reissner–Nördstrom (charged), and Kerr–Newman (spinning and charged). In practice, however, it is unlikely that black holes have any charge; the most likely kind is the Kerr black hole.

What about black hole candidates? Do we have any good ones? We already saw that we do have one, namely Cyg X-1. But we also have others. One of the best is in the Large Magellanic Cloud; it is called LMC X-3. Like Cyg X-1, it is a binary system that emits X rays. Calculations show that it has a mass of about seven solar masses—considerably more than the three solar masses needed. A second source in the Large Magellanic Cloud, LMC X-1, may also be a black hole. It is an X-ray source, but in general it is not as good a candidate as LMC X-3. Another good candidate is called SS433. It caused considerable excitement when scientists discovered in 1978 that it was an X ray source. Observations soon showed that it had three sets of emission lines in its spectrum, each of them acting differently. Both red- and blueshifts were observed. Furthermore, the velocities were incredible: 50,000 km/sec. Astronomers now believe that this system may contain a black hole.

Cir X-1, in the constellation Circinius, was also at one time considered to be a good candidate. It is an X-ray source, but recent information indicates it is probably a neutron star. Besides individual stars, astronomers have a large array of candidates of another type: radio galaxies and quasars. A giant black hole may reside in the core of these objects, their radio energy being generated as it pulls in nearby stars. When nearby stars are no longer available the galaxy will settle down and become an ordinary galaxy like ours. Thus, it is even possible that our galaxy has a black hole at its center.

BLACK HOLES AS DARK MATTER

Now that we know a little about black holes let's come back to our question: Are they a good candidate for the dark matter? Let's begin with stellar-collapse black holes. As I mentioned earlier they arise when massive stars collapse. The final mass only has to be greater than 3 solar masses, but the star loses mass during its life, and in the final collapse. It's therefore likely that a star of about 8 solar masses would be needed to create a black hole. And although there are stars as massive as 50 solar masses, there are very few even as massive as 8 solar masses. It seems unlikely therefore that there would be enough to make stellar-collapse black holes a good candidate.

Furthermore, there's another problem. Stellar-collapse black holes arise from stars—and stars are made up predominantly of baryons. As I mentioned earlier, arguments based on nucleosynthesis show that there are not enough baryons in the universe to close it, or enough to give a substantial amount of dark matter.

What about primordial black holes? They are quite a different story. One of the major reasons is that many of them formed before nucleosynthesis and the above argument doesn't apply to them. Primordial black holes are different from stellar-collapse black holes in that they can range all the way from tiny ones up to extremely massive ones in the cores of galaxies. Let's begin with the gigantic ones. At first sight they look like particularly good candidates, but when the details are examined we find several problems. First, if there were many massive black holes in the cores of galaxies, the galaxies in clusters would have much higher velocities than they do. Furthermore, if they were not in their cores there would be evidence of distortion in the galaxies in clusters. Sidney Van den Burgh has looked into this for the case of the Virgo cluster and found no evidence for distortion due to black holes. Another argument against them is that we would see their "signature" as an inhomogeneity in the cosmic background radiation—and we don't. All of these seem to argue strongly against supermassive black holes as good candidates.

What about smaller ones? We obviously have to eliminate all those less than 10^{15} g, as we saw earlier that they have all exploded and evaporated off into space. David Schramm became interested in this problem in 1981 while he was at the University of Utah as an adjunct professor. Also at the University of Utah was Richard Price, an expert in black holes. Together with a graduate student, Katherine Freeze, they decided to look at small black holes as a possible candidate for the dark matter. They soon discovered that the best candidates were in the range 10^{15} to 10^{33} g, now called planetary-mass black holes (PMBH) because they are roughly the mass of a planet.

Richard Price.

I asked Price about the project. "Schramm was the inspiration for it," he said. "I had done some work with Jim Ipser at the University of Florida on the question of black hole luminosity . . . so I was familiar with the literature and thinking in relation to whether small black holes could constitute some of the dark matter. Dave knew I had done work in this area and he suggested the collaboration. It worked out well. We put constraints on what could be constituents of the dark matter, but we really weren't about to start beating a drum for this explanation."

According to their paper, which was published in 1983, primordial black holes in the mass range 10^{15} to 10^{33} g are good candidates for dark matter. Price said he feels that their arguments still stand. So, while most black holes are not good candidates, planetary-mass ones may be.

CHAPTER 8

Neutrinos with Mass

Another of our candidates is the neutrino—a ghostly particle with no charge or spin, and little or no mass. In fact, it's probably fair to say it has little of anything. It therefore might seem strange that we were ever able to discover it. The experiment in which it was first detected was, indeed, a credit to the scientists who performed it—it was probably the most difficult experiment ever performed to that time.

But is the neutrino a good candidate for dark matter? Before 1980 there were strong indications that it had zero rest mass, and it was not taken seriously. Of course, even with zero rest mass, particles contribute some mass because of their energy (according to the theory of relativity energy is equivalent to mass). And since we know roughly how many neutrinos there are in the universe we can calculate their total contribution to its mass. As a massless particle they contribute little. But when it was discovered in 1980 that they might have a small mass, cosmologists began to take a second look at them. There are so many of them in the universe that even with a tiny mass they would make a substantial contribution, and would be a serious candidate for dark matter.

THE DISCOVERY

To understand the contribution of the neutrino it is best to begin with its discovery. Near the beginning of this century scientists detected what they thought were three types of radiation emanating from radioactive materials. It was later shown, however, that only one of the three types was actually radiation. For lack of a better name, perhaps, they were called alpha (α), beta (β) and gamma (γ) rays. The alpha rays turned out to be helium nuclei, the beta rays were shown to be electrons, and the gamma rays, electromagnetic radiation.

We will be concerned only with the beta particles, or more exactly, a process involving beta particles called beta decay. Soon after the discovery of beta decay scientists noticed a serious problem. Although the neutron had not yet been discovered at this time, the best known beta decay today is neutron decay so I'll explain the problem in terms of it. If a neutron is pulled from an atom it beta decays to a proton and an electron in about 12 minutes. If you carefully weigh all three of the particles you find that the neutron weighs slightly more than the combined weight of the proton and electron. But, of course, energy must be conserved in any reaction, so the electron must carry off the excess energy as it is released in the decay. But scientists noticed that it wasn't. Electrons were coming off with a range of energies: everything from zero up to the amount they should have.

What was going on? What was happening to the missing energy? Scientists were stumped. It almost seemed as if the conservation of energy was being violated. And a number of prominent scientists even went as far as suggesting that this might be the case. But conservation of energy is one of the pillars of physics; it would be blasphemy to suggest it was broken—and they knew it. A further surprise came when scientists measured the momentum (mass multiplied by speed) of the particles. Momentum must also be conserved—and it wasn't. Alpha decay, which is quite similar to beta decay, obeyed conservation laws; why didn't beta decay? It was a mystery that seemed to go to the roots of physics. Was there any way around it? In 1930 Wolfgang Pauli pointed out that there was: he postulated that a third particle—invisible to us—carried off the missing energy and momentum.

By the time he made this suggestion Pauli had already built a solid reputation in physics. At 21 he wrote a technical monograph on the theory of relativity that soon became a classic. And at 25 he formulated the quantum mechanical Exclusion Principle that forbids any two electrons with exactly the same properties from sharing a certain small region of space.

At this stage in his life Pauli was still making important contributions, but in later life he became somewhat of an eccentric. Although he read virtually everything that was published he made few important contributions; his major role was that of a critic. And a sharp, acid-tongued critic he was. Almost anyone who came near him felt his sting.

A professor was giving a talk one day with Pauli sitting in the first row. Pauli criticized and interrupted him so often he finally stopped, looked him straight in the eye, and said, "Dr. Pauli, I can't help it if I don't think as fast as you." Pauli stood up and shouted back, "I don't hold it against you that you don't think as fast as me. What I hold against you is that you publish faster than you think." With that the professor sat down and the talk came to an end.

The meeting at which the "invisible" particle suggestion was made took place in Tübingen in December 1930. Pauli was invited but it conflicted with a dance tournament he wanted to enter in Zurich so he sent a letter to the directors of the conference, Hans Geiger and Lise Meitner. "Dear Radioactive Ladies and Gentlemen," it began. He went on to describe the new particle of beta decay, calling it a neutron because it had no charge. It's hard to say how serious the participants took the suggestion, but it is evident that Pauli himself was not sure of it, as he ended his letter with, "Nothing ventured. Nothing won."

But Pauli didn't let the idea die. A year later he made the suggestion at a meeting of the American Physical Society in California. But still he published nothing. In fact, it seems as if he didn't even make detailed calculations of its properties. And, as it turned out, the name he selected for it was short-lived. In 1932 James Chadwick of England discovered the neutral particle

$$R_{IK} + \tfrac{1}{2} g_{IK} R = -KT_{IK}$$
$$R_{IK} + \tfrac{1}{2} g_{IK} R - g_{IK} \lambda = -KT_{IK}$$

"Dr. Pauli, I can't help it if I don't think as fast as you."

in the nucleus of the atom and, unaware of Pauli's particle, he named it the neutron. For a while, therefore, we had two particles with the same name. But Enrico Fermi of Italy came to the rescue: he renamed Pauli's particle the neutrino (which means "little neutron" in Italian).

Fermi, who would later achieve the first sustained fission reaction, became intrigued with the neutrino, and in 1933 he used it to develop a theory of beta decay. He described his theory to three of his colleagues one evening after a hard day's skiing in the Alps, saying he was convinced it was the best thing he had ever done. But oddly enough when he sent it to *Nature* for publication it was turned down. "Too speculative," said the editor. Fermi therefore submitted it to a relatively obscure Italian journal where it was accepted. It is perhaps an understatement to say it was an amazing theory; it was so good it stood virtually intact for over 30 years.

It might seem strange that no one ran out immediately to look for the neutrino. Fermi showed, however, that even if they had, they wouldn't have found it; according to his calculations it was almost impossible to detect. If you built a lead shield from here to the nearest star most neutrinos would pass right through it without interacting in any way. The extremely small probability of detecting it prompted Pauli to say, "I have done the worst thing for a theoretical physicist. I have invented something which cannot be detected experimentally."

And for many years it seemed hopeless. Fermi had plotted the energy distribution of the electron under the assumption it had zero mass, and also for the case where it had a small mass. And it had generally been decided that it had zero mass. Furthermore, with no charge its track would not be visible in a bubble (or cloud) chamber. How would anyone go about detecting it?

Oddly enough, it was detected. It took a few years, but in the mid-1950s Fredrick Reines and Clyde Cowan of Los Alamos National Laboratory set up an experiment that allowed them to "see" a neutrino (indirectly) for the first time. They knew that it would be extremely difficult to detect, but if they had enough— billions of them—they felt they might be able to detect a few a day. Nuclear reactors, it turns out, give off about 10^{13} neutrinos per square centimeter per second. To be perfectly accurate they actually give off antineutrinos (the antiparticle of the neutrino) but if antineutrinos exist so do neutrinos. In case you're not familiar with antiparticles, they have the same mass as their corresponding particle, but usually have an opposite charge. When a particle and its antiparticle collide they annihilate one another with the release of photons.

Reines and Cowan selected a reactor along the Savannah River in South Carolina. Beside it they built their detector, which consisted of 12 tons of water sandwiched in compartments between photoelectric cells. According to their calculations most of the 10^{13} antineutrinos would pass right through it, but a tiny fraction—about two or three an hour—would react with their apparatus. At first they had problems, but they added more shielding. And finally they saw what they were looking for: evidence of antineutrinos.

COMPLICATIONS

With the detection of the neutrino, scientists began to feel they were finally starting to understand the strange particle. But more surprises were soon to come. The neutrino that had been detected was associated with the electron; we now refer to it as the electron neutrino. But in 1936 Carl Anderson of the California Institute of Technology had detected what appeared to be a heavy electron—now called a muon. It seemed so out of place that the Nobel Prize winner I. I. Rabi, upon being told about it in the early 1940s, said, "Who ordered that?" Nobody was quite sure why we needed a heavy electron. It was so much like the electron, though, that scientists soon began to wonder if it also had a neutrino—perhaps different from the electron neutrino. An experimental check was performed at Brookhaven in 1962 by Leon Ledermann and Melvin Schwartz of Columbia University. And sure enough, it had a neutrino—different from the electron neutrino (it would not interact with the electron). We're still not sure exactly how the electron and muon neutrinos differ, but they are different. Furthermore, in 1977, another, even heavier electron was discovered; it was called tau. And we assume it has a neutrino—different from the other two.

How are these three particles related? One possibility was suggested in the late 1950s by the Soviet physicist Bruno Pontecorvo. Using quantum theory he showed that if each of the neutrinos had a mass, and the masses were different, they could "oscillate," or change identities as they move through space. An electron neutrino could become a muon neutrino, and so on. The idea was interesting but no one paid much attention to it at the time.

In the late 1960s Ray Davis of Brookhaven began looking into the possibility of detecting neutrinos from the sun. It was well known that tremendous numbers were generated in the core of the sun, and literally all of them escaped immediately. Davis discovered a reaction involving chlorine and argon that would allow him to detect them. He would need considerable shielding to stop cosmic rays from interfering with the experiment so he set it up in a deep mine at Lead, South Dakota. Basically, it consisted of a huge tank of cleaning fluid (which contained chlorine). When neutrinos reacted with the chlorine atoms in the cleaning fluid they changed them to argon atoms. The argon atoms could then easily be swept from the tank and counted.

The experiment was a success. Argon atoms were found, indicating the presence of neutrinos. But surprisingly, only about one-third the number of neutrinos expected were detected. What happened to the other two-thirds? Over the years many explanations have been given. But only one is now taken seriously: it is associated with the oscillations mentioned above. Davis's experiment can only detect electron neutrinos, and the number he expects is based on calculations of the number of electron neutrinos generated in the sun. But if they oscillate as they travel to us, there will be a mixture of all three types by the time they get here. Only one-third will be electron neutrinos; two-thirds will be types he cannot detect.

To check on this idea Davis hopes to build a detector that will detect neutrinos of all types. But gallium, a rather rare metal, is needed for this detector, and there is a problem with its cost and availability. In the early 1970s, when the experiment was first conceived, it was hopelessly out of range—all the gallium that had ever been produced in the world would be needed. But gallium is now being used more and more in electronics, and with increasing amounts being produced, the cost is coming down. Still, about 50 tons of gallium are needed—at a cost of about 25 million dollars. And Davis has had trouble coming up with such a large amount of money, so it may be a while before the experiment is performed.

DOES THE NEUTRINO HAVE MASS?

Not only does Pontecorvo's oscillation idea help solve Davis's problem, but it also means that the neutrino has mass. It would have to be a tiny mass; Fermi showed years ago it could not be greater than about a ten-thousandth that of the electron. This would make it by far the lightest known particle (aside from the zero mass photon). But even with a tiny mass it could be extremely important. Neutrinos were generated in the big bang explosion; according to the big bang theory they now outnumber ordinary particles (baryons) by a factor of 10^8. This makes them as abundant as photons (about 450 per cubic centimeter), which means that even if they have a tiny mass they can make up most of the mass of the universe. They could, indeed, be the missing mass.

Also, neutrinos could explain the massive halos around galaxies. If they have mass they will not travel at the speed of light, and could travel as slow as a few kilometers per second. If so they would be captured by galaxies and clusters of galaxies. And if their mass is in the right range they could account for their halos.

The first group to look for neutrino mass was one composed of Fredrick Reines, Henry Sobel, and Elaine Pasierb of the University of California. Actually, they weren't trying to measure the mass of the neutrino directly; they were looking for oscillations, which would indicate it had mass. It was a difficult experiment, and even after they announced in 1980 that they

Weighing neutrinos: "Ten billion-billion-billion more and we'll reach the sensitivity of the scales."

had detected oscillations, Reines expressed his concern, saying, "It will need verification." But verification did not come. And the experiment would likely have been forgotten, except that shortly after it was performed a Soviet team, headed by E. F. Tret'yakov, announced that they had detected neutrino mass directly in the beta decay of tritium, a heavy form of hydrogen.

This sent the world of science into a dither. Did the neutrino have mass or didn't it? Experiments were set up to duplicate the Soviet one. And it too was not verified.

I asked Gerhard Börner of the Max Planck Institute in Germany for his opinion on the matter. He is working on dark matter and galaxy formation. "The Russians have repeated their experiment," he said. "They now get about 30 electron volts [the energy equivalent of the mass]." Then he shook his head. "But I'm not convinced their experiment is valid. They only have a resolution of 30 electron volts, which means their measurement is consistent with 0 electron volts, or zero mass. The experiment was repeated by a group at Zurich. They got an upper limit of 14 electron volts, but it was also consistent with 0 electron volts, or zero mass. The problem just hasn't been resolved yet."

Gary Steigman of Ohio State University says, "The Russians are willing to take their uncertainty down to about 20 electron volts. But the Zurich experiment doesn't agree with it. So if you're an optimist you would say that the Russian experiment still hasn't been excluded; if you're a pessimist you would say there is accumulating evidence that the Russians are not correct. At the present time I think it's fair to say that there's no good evidence for neutrino mass. It wouldn't surprise me, though, if neutrinos do have mass and the heaviest ones are only a fraction of an electron volt, in which case, as far as cosmology is concerned, the mass wouldn't play a significant role. But it could be used to explain the solar neutrino problem [Davis's experiment]."

Things looked bleak for neutrino mass, but on February 23, 1987, astronomers received a gift from heaven: a supernova (exploding star) in the Large Magellanic Cloud. It caused a sensation. Within a day every major observatory in the southern hemisphere (it could only be seen there) had their telescopes trained on it. Its spectrum was quickly obtained using the IUE satellite. But most important for neutrino astronomy, neutrinos were detected at four different sites. And since neutrinos with mass don't travel at the speed of light, a measure of their time delay (or dispersion) compared to the photons would give an indication of their mass. Eleven events (pulses) were received by the Kamiokande II detector in Japan and eight at the IMB detector at Brookhaven. Examination of these events allowed scientists to put limits on the mass.

First estimates came from John Bahcall and Sheldon Glashow, who found a mass about half that of the Soviet experiment. Adam Burrows of the University of Arizona made a more detailed analysis that agreed with their result. Later, Edward Kolb, Albert Stebbins, and Michael Turner of Fermilab pointed out that the limits obtained from the supernova were not really any better than present laboratory limits. "There has been a lot of confusion over this supernova," said Börner. "That type of star should not have exploded." Astronomers have thought for years that only red giants explode. But the star associated with this supernova was a blue giant. There are still problems but theorists now feel they have a satisfactory model of the explosion.

MASSIVE NEUTRINOS AS DARK MATTER

On the basis of what we now know let's weigh the evidence. Is the neutrino a good candidate for the dark matter around galaxies, or for the missing mass of the universe? The case for it has been getting weaker in recent years but it still has much going for it. Compared to the particles we will be talking about in the next chapter—axions, supersymmetric particles, and so on—it has a distinct advantage. We know it exists. But if it exists with zero rest mass it cannot be considered seriously. To be considered seriously it not only has to have a mass, but the mass must be in the proper range.

If the mass is, indeed, in the right range, neutrinos could provide the dark matter around galaxies and clusters. Gary Steigmann was one of the first to point this out. "David Schramm and I won the gravity prize several years ago for showing that neutrinos would cluster on a large scale but not on a small scale," he said. "This seemed to be in agreement with the data on how the dark matter was distributed in the universe. That was the first shot at the problem, and many people have now looked at the possibility in great detail. And the neutrino now seems to be in trouble . . . at least in the context of calculations done so far."

Another advantage of the neutrino is that we don't need to worry about the restrictions of nucleosynthesis. They apply only to baryons, and neutrinos are not baryons. Furthermore, neutrino mass is consistent with Grand Unified Theories; such theories don't require it, but at the same time they don't forbid it.

Despite its advantages, however, the neutrino has serious problems. One is related to its velocity. If its mass is great enough to close the universe its velocity would only be about 10 km/sec, which is so slow that neutrinos would not only be captured by galaxies, they would fall right through to their cores. In particular, they would not be moving fast enough to keep them in an extended halo around the galaxy where they would have to be to solve the stability problem of galaxies.

Furthermore, if they had sufficient velocity to keep them in a halo, their mass would not be large enough to close the universe. Extending this argument to halos around clusters of galaxies makes things even worse. In addition, although it is possible that neutrinos make up the dark matter in galaxies and clusters, astronomers have shown that they could not be the dark matter associated with dwarf galaxies, and we know that dwarf galaxies contain dark matter.

Another problem, of course, is that their mass has not yet been verified. The 1987 supernova was helpful; it seemed to confirm laboratory estimates. But if they are accepted for the electron neutrino (the type given off in supernova explosions) it would have to be eliminated as a candidate; it is too light. This, of course, still leaves the muon and tau neutrinos.

Some of the most serious difficulties for neutrinos, however, relate to the formation of galaxies and large-scale structure in the universe. If the initial cloud consisted mainly of neutrinos, a study of the fluctuations that caused the galaxies and the large-scale structure, puts restrictions on the mass of the neutrino. As we will see in a later chapter most models of structure formulation based on neutrinos are now in trouble.

So, although a few continue to argue that the neutrino still has possibilities as a viable candidate for the dark matter, most scientists working in the area are convinced that interest in it will eventually fade away.

CHAPTER 9

Magnetic Monopoles and Other Exotic Particles

Because of the restriction from nucleosynthesis that the dark matter cannot be in the form of baryons, scientists have, in recent years, turned toward what are called "exotic particles." They are referred to as exotic because we're not sure they exist—in short, they exist only on paper. I'll begin with what is called the magnetic monopole. There has been a tremendous amount of interest in it the last few years, not only because it would be a good candidate for dark matter, but also because scientists have predicted that it should exist, and so far—despite extensive searches—it hasn't been found.

MAGNETIC MONOPOLES

To understand what a magnetic monopole is we must start with magnetism. You may remember sprinkling iron filings on a sheet over a bar magnet in school and watching the pattern that resulted. It clearly showed that there is a field around a magnet. According to present convention, the field lines come out of the north pole of the magnet, curve around it, and enter again at the south pole. It is also easy to see that individual lines do not cross. Although the phenomenon of magnetism has been known for centuries (it was even noted in early Greek writings, about 500 B.C.) it was not until 1819 that it was discovered to be associated with electricity. In that year Hans Christian Oersted was giving a classroom demonstration on electrical currents when, by accident, he left a compass near a wire he was working with. To his surprise he noticed the needle jumped when he passed a current through the wire. Looking closer he saw that the needle pointed not in the direction of the current (or opposite it), but perpendicular to it. He then reversed the direction of the current and found that the needle reversed direction. Astounded by the phenomenon he stayed after class and continued to experiment with it. The following year he announced his discovery, and within months all of Europe was experimenting with electricity and magnetism.

But if electricity could generate magnetism, it was natural to ask if magnetism could generate electricity. The English physicist Michael Faraday tackled the problem, and showed that a changing magnetic field could, indeed, produce a current. Using a loop of wire and an instrument to measure current, he moved a magnet through the loop and noticed that it produced a current.

Thus, a changing magnetic field produced a current (and with this current there was an associated electric field), and a changing electric field produced a magnetic field. When Faraday made his discovery, however, the idea of a field was unknown. Indeed, the concept came from him. He imagined invisible "lines of force" around a magnet, but because he didn't have a mathematical background he was unable to express his idea mathematically. This was done by the British mathematical physicist Clerk Maxwell. Maxwell wrote down four equations that showed the relationship between electricity and magnetism. There was a symmetry between them—with one exception. Although there was an electric charge that gave rise to an electric field, there was no corresponding magnetic charge. In other words, a single north or south pole did not appear to exist. Maxwell wondered about the lack of symmetry, and many others have since. It seemed reasonable that a single magnetic charge—a magnetic monopole—should exist.

To see why, consider what is called an electric dipole. We can make one up by placing a positive charge on one end of an insulating rod, and a negative charge on the other end. The electric field lines around the dipole will resemble the magnetic field lines around a magnet. If we now cut the insulating rod and separate the charges we get two monopoles (one positive and one negative)—each with an electric field around it.

Suppose, now, that we perform the same experiment with a magnetic dipole (a bar magnet is a dipole). As before let's cut the

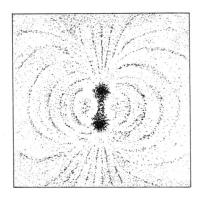

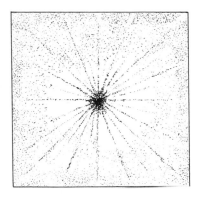

Upper: Dipole. Lower: Monopole.

magnet in half and separate the two halves. If we do we find we don't get two monopoles as in the above case; we still have a dipole field—in other words, each half of the magnet has a north and a south pole. If we cut each of these smaller magnets in half again we get even smaller magnets each with a north and a south pole. In short, it appears impossible to make up a monopole.

But do magnetic monopoles exist in nature? Paul Dirac of Cambridge University showed mathematically that they should. Dirac was actually looking at another problem when he made the discovery. One of the pioneers of quantum theory, Dirac was confused about the "quantization" of charge. Why, he asked himself, does charge come only in integral amounts of the electronic charge? In 1931 he was able to show that this could be explained if magnetic monopoles existed.

Scientists soon became excited about Dirac's prediction. If magnetic monopoles existed we should be able to find them. But where would we look? Dirac showed that its magnetic charge would be 69 times as great as the electronic charge, but his theory did not predict the monopole's mass. And this made it difficult for searchers. Furthermore, there was little to guide them as to how fast monopoles would be traveling. Most believed, however, that they would be stopped by Earth, so many of the searches were made underground.

There was, however, a difficulty with Dirac's monopole: it had a tail, or string, attached to it. No one knew what the significance of this tail was—and it seemed so unreal. Certainly, none of the known particles had tails. Despite the difficulty, scientists undertook extensive searches; they searched the bottom of oceans, the tops of mountains, cosmic rays, and accelerator tracks. But they found nothing. Then in 1969 the first rocks were brought back from the moon. They were over 4 billion years old, and if monopoles lodged in anything, they should be here. Luis Alvarez of the University of California built a detector for checking them. It consisted of a loop of wire that he passed back and forth across the rocks. If a monopole passed through the loop it would generate a current. But again none were found.

By the early 1970s interest in magnetic monopoles had begun to wane. Information on its properties was lacking, and most of the likely places had been searched. But about this time interest in another area was building. Scientists had known for years that there were four basic fields of nature. The gravitational field is familiar to everyone, and we just talked about the electric and magnetic fields, or electromagnetic field. Besides these two fields, though, there are two nuclear fields. One holds the particles of the nucleus (protons and neutrons) together and the other is associated with radioactive decay. We refer to them as the strong and weak nuclear fields.

What is the relationship between these four fields? Einstein believed that two of them, the electromagnetic and gravitational, were just different forms of the same field. In other words, he was convinced a mathematical theory could be found that would "unify" them. He struggled for over 30 years to find such a theory, but failed. In 1967, however, Steven Weinberg and, independently, Abdus Salam showed that the electromagnetic field could be unified with the weak nuclear field. While not a perfect unification, it was a significant breakthrough; they were later awarded the Nobel Prize for the discovery.

But if two of the fields could be unified, could the other two be brought into the fold so that all fields were unified? Scientists started with the strong nuclear field, and in 1974 Howard Georgi and Sheldon Glashow of Harvard University published a theory indicating a possible unification. Other similar theories soon followed. We now refer to such attempts as grand unified theories (GUTs).

The real breakthrough for monopole hunters, however, came in 1974 when Gerard 't Hooft of Holland and, independently, H. Polyakov of the USSR showed that grand unified theory predicted the existence of magnetic monopoles. Indeed, they went much further than Dirac in that they were able to predict its mass: it would be 10 million billion billion times as heavy as the proton. No wonder no one had found it. The only place particles this massive could have been produced is the early universe. In fact, they would have to have been produced before about 10^{-34} second; only then was the energy high enough. If they were produced in large numbers and are uniformly spread throughout the universe now, it seems likely that we would be able to detect them. Early predictions from grand unified theory, if fact, indicated that they would be as common as protons. But strangely, this presented a serious problem: with their large mass they would have stopped the expansion of the universe; in fact, it would have collapsed in on itself in just 10,000 years.

Something was wrong! The prediction couldn't possibly be correct. There were too many monopoles; their number had to have been diluted. Blas Cabrera of Stanford University began looking into the problem in the early 1980s. He determined how many it would take to barely close the universe, and found that with the appropriate equipment he would be able to detect about two a year. This seemed like a large enough number so he went ahead and built an apparatus similar to the one Alvarez

Blas Cabrera.

had used several years earlier. But Cabrera took many more precautions. Like Alvarez he used a loop of wire, assuming that if a monopole passed through it, a small current would flow. The magnitude of the current would not depend on the mass of the monopole, but it would depend on the charge. Since the charge was known, he could predict the magnitude of the current.

The first thing Cabrera had to do was make sure the region around the loop was field-free—in particular, Earth's magnetic field had to be eliminated. To do this he used a phenomenon known as superconductivity. It is well known that when certain materials are cooled to a few degrees above absolute zero they lose their electrical resistance and conduct electricity easily. In fact, once a current is set up in a loop it continues indefinitely. Cabrera cooled a deflated lead foil balloon to superconducting temperatures, then inflated it. This pushed any magnetic field around it away, and left the region inside the balloon field-free. He placed his loop in the center of this field-free region.

Because the expected current was so small a special device called a SQUID (superconducting quantum interference device)

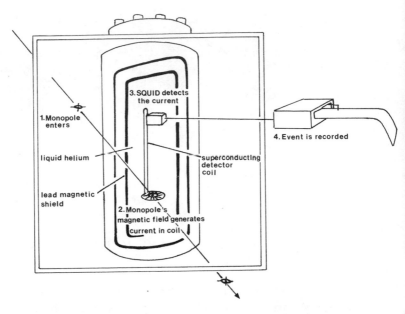

Schematic of Cabrera's apparatus.

had to be used to detect it. The signal from the SQUID was fed to a strip recorder; if a sudden current flowed in the loop it would show up as a blip, or pulse, on the recorder.

Cabrera and his students built the apparatus and checked it thoroughly to see if pulses were generated by any outside interference. Once they were satisfied it was operating properly, they waited. For six months they waited with nothing showing up. Then, on Monday, February 15, 1982, Cabrera came to work, going directly to his lab to check on his experiment. To his surprise it showed a pulse of about the right size to be a magnetic monopole. He was excited, but cautious. He knew the consequences of a premature announcement. Only a few years earlier (1975) Buford Price, at Berkeley, had called a press conference to announce his discovery of a magnetic monopole. The news spread rapidly and generated considerable excitement. Then it was discovered that Price had made a mistake: what he saw couldn't have been a magnetic monopole. Cabrera didn't want this to happen to him. He felt some comfort, however, knowing that the pulse was about the right size.

Over the next few days he thoroughly checked his apparatus to see if it was an instrument error—perhaps a loose wire. But he found no problems. He also checked to see if there had been any nearby earthquakes. There were none. And there were no nearby explosions of any kind. He began to feel confident that it was, indeed, a monopole, and started to write a short, but cautious report of the event for publication in *Physical Review Letters*.

By now, however, news of his "discovery" had leaked out. And it was soon picked up by the press. Numerous reporters came to interview him, but he wouldn't talk to any of them until he was sure it was a monopole. And, at this stage, other scientists were still skeptical. As far as most of them were concerned a single pulse didn't prove anything.

Other experiments were soon set up in an effort to confirm the result, and in 1984 another group, working under S. N. Anderson at the University of Washington, reported they had detected five monopoles. Cabrera also built a second detector—a larger, more accurate one.

But as time passed no more monopoles were detected, and the skepticism remained. One of the reasons for the doubt was that E. N. Parker of the University of Chicago had shown in 1970 that monopoles could destroy the magnetic field of our galaxy. According to Parker, if two monopoles per year were detected, the magnetic field of our galaxy could not exist. Yet we know it does—we can measure it. Parker concludes that monopoles should be at least 10,000 times less common than Cabrera's experiment indicates. (With a larger detector than Cabrera's, however, they could still be observed at the same rate.)

Ira Wasserman of Cornell University, however, does not agree with Parker. He argues that any magnetic monopoles captured by galaxies would enhance their field rather than destroy it. Both his and Parker's arguments, however, are in conflict with the prediction of grand unified theory. But, as we saw earlier, if this prediction is correct, monopoles had to be diluted in the early universe. How? In 1980 Alan Guth, now of MIT, began looking into this problem. And what he discovered was that the universe appeared to have undergone a sudden "inflation" shortly after it began. Although Guth devised the theory in an effort to show that the monopoles would have been diluted, he found that it solved several other problems of the big bang theory—yet, strangely, it did not solve the monopole problem. Furthermore, his theory had another difficulty: he was unable to bring the inflation to a smooth finish. In 1981, however, Paul Steinhardt and Andreas Albrecht of the University of Pennsylvania, and Andrei Linde of the USSR independently showed that, with a slight modification, Guth's inflation could be brought to a smooth ending. And in the process they showed that few monopoles would have been generated.

Interestingly, although they solved the problem of the overabundance of monopoles, they created a new one. According to their theory, now called "new inflation," only one or two monopoles exist in the entire universe. If true, it would hardly be worthwhile searching for them. But there are still problems with inflation theory, and it is not yet universally accepted. Most scientists working in the area are confident that some sort of inflation occurred, but it may not have left only a few monopoles. There are, in fact, versions of inflation in which large numbers of monopoles are generated. In any event, inflation has not discouraged monopole hunters.

Since Cabrera's original experiment many others have been set up. A group referred to as IMB (consisting of physicists from the University of Illinois, the University of Michigan, Brookhaven, and others) has set up two experiments similar to Cabrera's. A group consisting of scientists from the University of Chicago and Fermilab are working on a similar design, as are groups at Kobe University in Japan and Imperial College in England.

Furthermore, the induction coil technique of Cabrera is not the only method of detecting monopoles. Because they are so heavy, it is assumed that they travel relatively slowly and their tracks will therefore be visible in an appropriate ionization chamber. Several groups are in the process of building such chambers. The largest is the Backsan underground detector in the USSR. It is a rectangular array of 3200 detectors. So far, though, no monopoles have been detected. Similar detectors are located in the Homestake gold mine near Lead, South Dakota, at Texas A&M University, and at Gran Sasso in Italy. Furthermore, there are other experiments using different types of detectors. But even with the large number of experiments that have been set up, no monopoles have been detected.

In many ways this is strange. The reason is that there are strong indications that monopoles have to exist. The symmetry between electricity and magnetism, Dirac's prediction, and the prediction from grand unified theory all argue for their existence.

MAGNETIC MONOPOLES AS DARK MATTER

Are magnetic monopoles a good candidate for dark matter? As far as their properties are concerned, they are. Monopoles would radiate little light and would be invisible. Furthermore, their predicted mass (10^{16} times the mass of the proton) is so large that relatively small numbers would make a significant contribution to the mass of the universe. Only about one monopole per 10^{16} protons would be needed to equal the amount of luminous matter in the universe.

The major difficulty with the monopole as a candidate is, of course, that we have not yet detected it. In addition, though, there are other problems. Dark matter is clumped in the halo of galaxies and clusters. Would monopoles tend to clump there? Michael Turner of the University of Chicago thinks it is unlikely. He points out that monopoles do not readily participate in collisions, and are therefore unlikely to be found in galactic halos. Furthermore, he says that the magnetic fields of galaxies would tend to accelerate passing monopoles rather than capture them. Coupling the above difficulties with the difficulties associated with the Parker limit and the predictions of inflation makes the magnetic monopole a relatively poor candidate at the present time. This would, of course, change dramatically if one was discovered.

I asked Gerhard Börner of the Max Planck Institute for his opinion. "They cannot be seriously considered at the present time," he said. "There are just too many problems. But I can imagine some fine tuning that might turn things around." Michael Turner, on the other hand, refers to monopoles as his "dark horse candidate." He feels they have possibilities.

SUPERSYMMETRIC PARTICLES

Several other exotic particle candidates come from a theory called supergravity. Earlier I talked about the attempt to bring the four forces of nature together into a unified theory. The grand unified theories are attempts to unify three of these fields, but the fourth, namely gravity, has proven particularly difficult to bring into the fold. With the formulation of supergravity in 1976, however, a way was found to include it. The foundations of supergravity were discovered by Soviet physicists in the early 1970s. Western scientists took little interest in the discovery until 1975 when Bruno Zumino, then at New York University, and Julius Wess of the University of Karlsruhe in Germany rediscovered the Soviet results and extended them; they called their theory "supersymmetry."

Wess completed his Ph.D. in 1957 at the University of Vienna. Between 1955 and 1966 he was an assistant professor at the Vienna Institute for Theoretical Physics. He is now at the University of Karlsruhe. Zumino received his doctorate in 1945 from the University of Rome. He held several positions before

Julius Wess.

becoming research associate at New York University in 1951. In 1982 he went to the University of California at Berkeley.

Wess and Zumino met at the University of Vienna while Zumino was there as a guest lecturer. "I went to his lectures and we eventually started to work together," said Wess. "The collaboration worked well." So well that when Zumino returned to New York University he wrote Wess and invited him to come to New York. Wess was delighted at the invitation and spent two years (1966–1968) in New York.

Their discovery of supersymmetry (a symmetry associated with the basic particles of nature) came at CERN in 1975. "Bruno and I were working on tensor theorems," said Wess. "But things weren't going well . . . we got frustrated and went to a seminar given by Sakata. That evening we went for a walk and talked about a possible symmetry associated with the things Sakata

Bruno Zumino.

had talked about." And within a short time they realized they were onto something important. "Once we realized what we had we were thrilled with the possibilities," said Wess.

A year later supersymmetry was extended by two different groups to include gravity. The resulting theory is now called supergravity. The importance of supergravity in relation to the dark matter problem is that several particles are predicted by it. According to the theory, corresponding to each known type of particle in our universe there is a "superpartner." These superpartners are given names such as photino, wino, squark, selectron, and gravitino. The photino is the superpartner of the photon, the squark, the superpartner of the quark, and so on.

One of the first of these particles to be seriously considered as a possible dark matter candidate was the gravitino, the superpartner of the graviton (the particle of the gravitational field). George Blumenthal and Joel Primack of the University of California at Santa Cruz and Heinz Pagels of the Rockefeller Institute published a paper showing that large numbers of gravitinos would have been produced in the early universe. According to their calculations, these gravitinos would be slow-moving and could easily form galaxy-sized clumps. Shortly after the paper was published, however, the gravitino was shown to be unstable. Attention was then focused on what appeared to be the only stable superparticle, the photino. Like gravitinos, photinos also move slowly and could produce galaxy-sized clumps.

I asked Wess what he thought about the photino as a candidate for dark matter. "It's natural to speculate that the dark matter might be photinos," he said. "I can't see any reason it couldn't be. But I'm not an expert on dark matter. Nobody can know for sure, though, until it is discovered." He then went on to say he was convinced it would be found when the superconducting supercollider was built.

Hans Bloemen and Joseph Silk of the University of California at Berkeley have shown, however, that photinos from the early universe could not account for all the dark matter in the halo of our galaxy. According to their calculations photinos could account for no more than about 15 percent, otherwise we would see more gamma rays than we do.

THE AXION AND OTHER CANDIDATES

In 1977 Roberto Peccei and Helen Quinn, who were then at Stanford University, introduced a new type of symmetry into grand unified theory in an attempt to overcome a problem that had developed. Shortly thereafter Frank Wilczek, now of the University of California at Santa Barbara, and Steven Weinberg of the University of Texas found that the Peccei–Quinn symmetry implied that a new type of particle had to exist; they called it the axion. It would be very light—a trillion times lighter than the electron—but according to their calculations large numbers of them should exist. Furthermore, they could clump around galaxies and clusters and would therefore be an excellent candidate for the dark matter.

Pierre Sikivie of the University of Florida pointed out that axions might be detected using a microwave cavity (cavity in which electromagnetic radiation resonates). Wilczek, John Moody of the University of California, and others have proposed refinements and alternate schemes.

The first experiment was set up by G. Galmini of Harvard University and his group. Using a germanium spectrometer, they were unable to detect any axions, but were able to place limits on its mass and abundance. I will talk about other experiments in a later chapter.

Besides the above candidates there are many others. Most are not considered seriously at the present time so I will only briefly mention them. First, there is shadow matter—an odd material that according to Michael Turner may have developed in the early universe. Shadow electrons and shadow protons could exist, but aside from their gravitational field, we could never detect them. There might even be shadow stars and shadow galaxies. Turner admits, though, that he does not take the idea too seriously any more.

Other possibilities are preons and rishons, the particles that some scientists believe make up quarks. And there are cosmic strings, superheavy strings that may have been responsible for the formation of galaxies. And finally a number of scientists have suggested that "nuggets" made up of quarks may be a good candidate.

It's hard to say how seriously we should take the exotic particles; many of them will no doubt turn out to be figments of our imagination. But as we will see in the next chapter some of them seem to be able to generate galaxies and large-scale structure better than most other particles.

CHAPTER 10

Galaxy Formation and Dark Matter

We have seen that there are many candidates for the dark matter, and we have looked in detail at some of them. In fact, we have even speculated as to which is best. But what we really need is something that can help us narrow down the candidates. And we do have something. It turns out that the structure of the universe—i.e., the form of the galaxies, clusters, and superclusters in it, and the spaces between them—depends on the dark matter.

According to the most widely accepted model, galaxies were created from the cloud that arose from the big bang. This cloud broke up as a result of fluctuations (ripples) within it. Once small regions of excess density developed they pulled other matter toward them and grew. Eventually they became galaxies. In the simplest version of the theory the cloud is assumed to be made up mainly of protons and neutrons (baryons), with about 90 percent of them making up the dark matter, the other 10 percent accounting for the visible matter.

The idea that fluctuations were present is now generally accepted, but the details of exactly how large-scale structure formed are another matter. The important thing, though, is whether or not the theory predicts what we see in the universe. Let's begin with the cosmic background radiation. This is radiation that was discovered in 1965 by Arno Penzias and Robert Wilson of Bell Labs. They found that the universe was filled with photons (radiation) at a temperature of approximately 3 K.

Where did these photons come from? Astronomers are now convinced that they came from the big bang. They are, in effect, an "echo" of the big bang.

According to the big bang theory the photons were "coupled" to the matter for the first 500,000 years. By this I mean they were continually being absorbed and emitted by the matter. They were, in effect, trapped by it. And because of this, stable atoms could not form; furthermore, any "lumps" that formed were quickly dispersed. The universe therefore remained generally uniform during this time. But eventually it cooled to a low enough temperature (3000 K) so that the photons could break away from the matter. And when they did, they flew freely off into space. Electrons and protons were then able to come together to form atoms.

The important point here is that we can now measure the "smoothness" of the decoupled photons (the cosmic background radiation). And it is exceedingly smooth—there are no variations in its uniformity greater than about 2.5 parts in 100,000. Since the matter was coupled to the radiation this means the fluctuations in it at the time of decoupling could not have been greater than this. It's easy to show, though, that galaxies could not have formed from such insignificant fluctuations. We therefore have a problem: How could galaxies form and still leave the cosmic background radiation so smooth.

One way around the problem is to assume that the dark matter was not baryons, but rather some type of particle that had little effect on the fluctuations. In this case about 1 to 10 percent of the particles present would be protons and neutrons, the rest would be dark matter of a different form. We have, indeed, seen that there are other candidates for the dark matter. And if they were generated in the early universe it is likely that they would have affected galaxy development. In this chapter we will see that astronomers now believe this is indeed the case. Before we look at the details, though, let's look briefly at the overall structure of the universe.

SURVEY OF THE UNIVERSE

As I just mentioned, any theory of galaxy formation must explain the uniformity of the cosmic background radiation, but it must also explain the large-scale structure of the universe. Interestingly, at one time astronomers were convinced that there was no large-scale structure; they believed that galaxies were uniformly distributed through the universe. But in the 1930s Fritz Zwicky showed that this wasn't true: many galaxies were in clusters. Large clusters such as the one in the constellation Coma Berenices were soon found. Then it was determined that our galaxy, the Milky Way, was part of a cluster of about 20 galaxies—now called the Local Cluster. Most galaxies, in fact, were shown to be in similar groups—some of these groups containing thousands of members.

But how far did this clustering go? Did clusters of clusters exist? In the 1950s the French astronomer Gerard de Vaucouleurs noticed that there was clustering on a higher scale, and he put forward the idea of superclusters. According to his observations our Local Cluster was part of a supercluster that was centered on a huge cluster in the direction of the constellation Virgo (called the Virgo cluster). But it took many years before his ideas were accepted.

For a while things seemed to be at a stalemate. What did the universe look like on the largest scale? Was it full of clusters and superclusters? To answer this astronomers obviously needed a detailed catalogue of the clusters. And in the late 1950s two surveys were begun, one by George Abell of UCLA, and one by Fritz Zwicky of Mt. Wilson. Abell found that over one-half of the galaxies in the sky were associated with clusters. Surprisingly, Zwicky found little evidence of clustering, but was eventually shown to be wrong. While the surveys did not settle the question they got others interested in the large-scale structure of the universe. Oddly enough, though, the next catalogue—one that played a very important role in the development of the sub-

A cluster of galaxies in the constellation Pisces. (Courtesy National Astronomy Observatories.)

ject—came about almost by accident. Donald Shane and Carl Wirtanen of Lick Observatory began photographing the stars in an ambitious project to determine how stars moved over the years. They photographed most of the stars in the northern hemisphere and planned on rephotographing them 20 years later. Shane noticed, however, that their photographs were dotted with background galaxies. It seemed a shame to ignore such a wealth of information; he therefore made a catalogue of them.

In the late 1970s the catalogue came to the attention of Jim Peebles of Princeton. He decided to use it to make a large-scale map of the sky. Along with several students he plotted a million galaxies and was amazed by what he saw: filamentary networks of knotlike galaxy clusters separated by voids. The universe was not uniform; it had a definite structure associated with it. But was the structure real? The plot was two-dimensional, and this meant we were seeing everything projected onto a single sheet. Galaxies were superimposed on one another. What was needed was a three-dimensional plot, and to get it a large number of redshifts would be needed. John Huchra of the Harvard–Smithsonian Center for Astrophysics and Marc Davis, now of the University of California, heeded the call and initiated a survey in 1976. Their results verified that there indeed were filaments and voids.

But more surprises were to come. Davis left the project and went to Berkeley, but Huchra along with Margaret Geller continued the survey to greater depths in the sky. And what they found amazed them. They expected filaments and voids, but found gigantic "bubbles" in the sky. Chains of clusters were strung along the outsides of the bubbles.

The importance of this large-scale structure is that astronomers believe its form depends on the dark matter in the universe. Or, more exactly, the structure depends on the initial fluctuations in the universe, and they in turn depend on the dark matter. And therefore structure and dark matter are closely linked. Anything we find out about one therefore helps in our understanding of the other.

MODELS OF STRUCTURE FORMATION

But how did the structure form? There are two models that attempt to explain it. In the early 1970s Yakov Zel'dovich of the Institute of Applied Mathematics in Moscow and several colleagues proposed what is now known as the "pancake" model. According to this model the original "clumps" of the universe were about the size of superclusters. When they collapsed they did not collapse at the same rate in all directions; in most cases the collapse occurred fastest along one direction creating a gigantic pancake. Clusters of galaxies formed along the intersection of these pancakes.

In this model superclusters formed first. Then came clusters, and finally, individual galaxies. The major prediction of the theory was that astronomers would find strings or chains of galaxies with huge voids between them. And this is, of course, what we do find. But there were two major problems with the theory. First, not enough time has elapsed since the big bang to produce galaxies. And second, there appeared to be no mechanism for creating the huge fluctuations that Zel'dovich postulated.

In 1980 several Soviets reported they had determined that the neutrino had a small mass. This was exactly what Zel'dovich and his team needed to overcome the problems of their theory. Neutrinos would decouple from the primordial cloud long before the photons. And for a considerable time after they decoupled their density would be generally uniform, but gradually as they cooled, they would begin to clump. And, according to Zel'dovich's calculations, these clumps would be about the size of superclusters. When ordinary matter finally decoupled it would be attracted to these clumps, and in the process the cosmic background radiation would be left uniform. In this theory the largest structures form first, with smaller structures forming later, and because of this scientists sometimes referred to it as the top-down theory.

For a while there was considerable interest in the theory. It could predict much of what was observed, but eventually it was found that even with the introduction of the massive neutrino, problems remained. The major one was that the theory did not permit galaxies to form properly.

About the same time as Zel'dovich put forward his theory, an alternate theory was proposed by Peebles of Princeton. He refers to his theory as the "bottom-up" theory. In it, the clumps that emerged at decoupling were galaxy-sized, and therefore the first things to form were galaxies. In time these galaxies attracted other galaxies and clusters formed. And finally clusters attracted other clusters and superclusters formed. But this theory also appeared to have a problem: not enough time has passed for superclusters to have formed.

How could we test these theories to see which is most in line with observation? One of the best ways would be to set up a computer program to simulate the evolution of the universe.

COMPUTER SIMULATIONS

Computer simulations are standard today, not only in astronomy, but throughout science. They are ideal for testing ideas. Scientists make a guess, then write a computer program that simulates the consequences of the guess. They then compare the results to observation, and if the agreement is close they assume the guess is correct. If not, they discard it and make a new guess.

In the case of galaxy formation, or more generally, the formation of structure in the universe, astronomers guess what the distribution of visible and dark matter is, and what it is composed of. They then run programs using the fundamental laws of motion. The motions of thousands, or in some cases, hundreds of thousands, of points representing clumps of matter are followed as they evolve in the early universe. Millions of years can be made to pass in the computer in seconds so astronomers can watch what happens to the clumps as they evolve over the entire history of the universe.

Simon White, who is now at the University of Arizona, was one of the first to become interested in computer simulations of this type. Born in Cornwall, England, White took his B.A. from Cambridge in mathematics. He then went to the University of Toronto in 1974 where he completed a master's degree in astronomy. It was here that he first became interested in the dark matter problem. And this interest continued when he went back to Cambridge for his Ph.D. His Ph.D. thesis focused on the dark matter that holds the Coma cluster together. He showed that it can't be bound to individual galaxies, but had to lie between them.

Simon White.

Upon completing his thesis he began a study of the effects of dark halos on the interaction of galaxies, showing that halos enhanced the collision and merger rates of galaxies by increasing their effective mass and size. His first taste of computer simulation came with this project.

In early 1981 he met Sergi Shandarin at a meeting in Sicily and learned of his computer models of the structure of the universe. "I was impressed by the potential of the methods," said White. "When I returned to the University of California at Berkeley I convinced Marc Davis and Carlos Frenk who were there to work with me on similar problems."

Marc Davis had just come to Berkeley. Born in Canton, Ohio, in 1947, he went to MIT, and then to Princeton for his Ph.D. I asked him how he got interested in astronomy. "I was interested in relativity . . . and space and time in high school," he said. "I read a lot about them, although I wasn't an amateur astronomer in that I didn't make telescopes or look at the stars a lot. In fact, I didn't even know the constellations that well. I was interested in the physics of space. I read Asimov and Gamow; they really motivated me to study astronomy and physics . . . and that's what I did."

After completing his Ph.D. Davis began working with John Huchra on a redshift survey of the sky. It was at this time that he first became interested in how the structure of the universe formed. "I saw all the large-scale structure [filaments and voids] and felt it was beyond what was explainable in our current models. So I felt it was time to try to improve the model, and to see if I could produce universe models that had any resemblance to what was observed." He therefore joined forces with White and Frenk, and together they began working on computer simulations of a neutrino-dominated universe.

Davis pointed out to me that before that time (early 1980s) it would have been impossible to do such a project. Cosmology and particle physics had just begun to merge, and for the first time astronomers were able to get realistic initial conditions (what the universe looked like at decoupling) for the universe.

Marc Davis.

"These new theories had the advantage that the initial conditions were prescribed. Before the early 1980s there was no
realistic way we could write down initial conditions," he said.

The first model they worked on is now referred to as the hot
dark matter model. They took the dark matter to be massive
neutrinos. When the first results came from the computer, Davis
and his colleagues were excited. Filaments and voids were
clearly visible. As they watched their model universe evolve—

millions of years clicking off in seconds—they saw the cloud of
particles become mottled. Clumps began to form, then filaments
and voids. But when they compared what they had to the real
universe the agreement was poor. Their model predicted too
many filaments.

George Efstathiou of Cambridge University then joined the
group. With his help they were able to significantly improve the
computer program. But the agreement was no better. "The more
we thought about the results, the more dubious we became that
the neutrino model could provide a quantitative fit to the real
world," said White. And in 1983 they published the first of
several papers suggesting that the massive neutrino universe
could not produce galaxies properly and was likely incorrect.

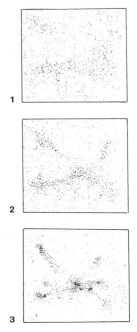

*The growth of structure in the universe according to the computer model of White, Frenk,
and Davis.*

"The computer simulations showed dramatically that structures formed very late in the universe, and that galaxies would really not exist," said Davis.

I asked him if he thought there was any way around the problem. "The only salvation for the massive neutrino model is if the initial conditions in the universe, the fluctuations, are not due to the neutrinos themselves, but are due to something else. People now talk about cosmic strings [massive, but very thin, strings that may have existed in the early universe] as seeds. The massive neutrinos are still there as the dark matter, but they cause no fluctuations." He paused. "But that's not an economical proposal . . . it requires several different kinds of dark matter. You've got to have strings and massive neutrinos. There's a principle in physics that says that an economical proposal is best—don't put in too many parameters. One at a time is what you should strive for." He went on to say that as far as he was concerned the simplest neutrino model was dead.

About the time the Berkeley group started their work, Adrian Melott, at the University of Pittsburgh, also began writing a computer program to simulate the formation of structure in the universe. "In the spring of 1980 I heard there was some evidence the neutrino had mass," said Melott. "This led me to wonder if neutrinos could be the missing mass of the universe." He decided to write a paper on the possibility. But many others had the same idea about the same time, and there was little notice of his suggestion. This did not deter him, however, and he soon began to write a computer program to see if his ideas were correct. The first question to be answered was: Could the neutrinos cluster in such a way as to allow dark halos around galaxies to form? He started with the simplest possible model: a one-dimensional one. Using it he was able to follow the collapse of the clumps soon after they appeared. At first they continued to expand with the universe, but eventually they stopped expanding and began to fall back on themselves. At this point it seemed as if a problem had developed: the average speed of the particles was too high to allow galaxies to form.

Looking closer, though, Melott saw that because there was a distribution of speeds, some particles (the low-speed ones) could settle down around the clumps and form galaxies with halos.

Melott then went on to two-dimensional models. In this program he put in small "lumps" to represent matter and found that filaments and voids formed. The next step was obviously the true representation of the universe: three-dimensional models. For these he collaborated with Joan Centrella of the University of Illinois. The program was now becoming exceedingly complex and it took several hours to run, even on the largest of computers—the CRAY. For months they worked together simulating and plotting three-dimensional neutrino-

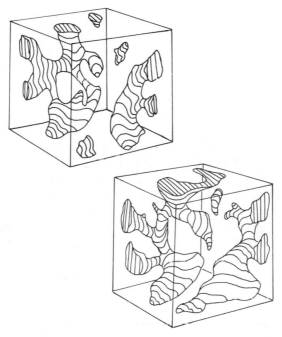

Computer models of voids and chains (Melott and Centrella).

filled universes. As they studied the graphs that resulted—representations of what the universe was doing over billions of years—they saw pancakes forming, just as Zel'dovich had predicted. And in some cases, when two axes collapsed, cigar-shaped regions formed. It appeared as if the voids were growing, squeezing the matter between them into long chains. These were chains of clusters.

But eventually, as in the case of the Berkeley group, Melott and Centrella began to realize the model wasn't going to work. Galaxies weren't forming properly. By 1983 they were convinced that something else was needed.

A COLD DARK MATTER UNIVERSE

In 1983 Davis, White, Frenk, and Efstathiou began to look beyond the neutrino model. But if neutrinos didn't work what else was there? As it turned out there was something else. In an earlier chapter I talked about supersymmetric particles (photinos, gravitinos, and so on), and other particles called axions. Collectively we refer to them as cold dark matter, using the word "cold" because they have low speeds as opposed to neutrinos, which have speeds near that of light.

Does it matter which of the cold dark matter particles we use? According to Davis it doesn't. "Our program is insensitive to the nature of the cold dark matter," he said. "It could be axions, photinos, or gravitinos . . . it doesn't matter."

The reason cold dark matter is such an ideal candidate is that it is coupled to the photons for only a brief time after the big bang. For most of the 500,000 years until ordinary matter decoupled it was free and therefore did not affect the photons (the cosmic background radiation).

Furthermore, the cold dark matter could have clumped and attracted baryons to the clumps after decoupling when the cosmic background radiation was already formed.

"Chains, voids . . . yes. But what I'd really like to know is what is beyond them."

Much to their delight, when Davis and his group sub-stituted cold dark matter for massive neutrinos in their program the agreement with observation was much better. A variety of structures were seen that could be identified with clusters and superclusters in the real universe. Furthermore, where the neutrino-dominated universe seemed to correspond to Zel'do-vich's pancake theory, the new simulations were more in agreement with Peebles's bottom-up theory. "This has turned into a very large and surprisingly successful project," said White. "We have completed five major papers on this topic so far and have shown that it is possible to make very detailed predictions for the properties of galaxies and galaxy clusters in a cold dark matter universe. Almost all of them agree reasonably well with observations."

Davis concurs. "I think cold dark matter models provide a coherent explanation of a lot of observations," he said. But he admits there are difficulties. "We're still not sure cold dark matter exists." (We saw in Chapter 9 that axions, photinos, and gravitinos exist only on paper.) White, in fact, says, "I rather doubt that the universe is actually filled with cold dark matter, but I suspect that we are uncovering a realistic picture of the way in which structure formed."

Melott and Centrella also soon switched their computer program over to cold dark matter. And they, like the Berkeley group, found that the results supported Peebles's theory. In short, they found that galaxies formed first, with clusters and superclusters coming later as galaxies were gravitationally attracted to one another. Similar results were obtained by Sandra Faber of Lick Observatory and George Blumenthal and Joel Primack of the University of California at Santa Barbara.

BIASED GALAXY FORMATION

The cold dark matter model of galaxy formation has been extremely successful, but as we saw it is not problem-free. A close examination of the data shows that it clumps too well. Furthermore, it predicts too much dark matter—50 times the amount of luminous matter. Another problem is related to the fact that most models are based on the assumption that the universe is flat—the prediction made by inflation theory. But if the universe is flat we have to ask: Where is all the cold dark matter that is needed to make it flat? Certainly it isn't tied up in the galaxies; we know that the matter we see there, even including that that is presumed to be present in halos, is far short of what is needed to close the universe. One way around this problem is to assume that we are not seeing most of the material of the universe. We know how much mass is associated with the visible, or "light," matter, but it may have little to do with the dark matter. If so, we must explain how this comes about. This means that our cold dark matter theory of galaxies must explain both the observed clustering and the dark matter problem. And in a new version, referred to as biased galaxy formation, it appears as if it does.

According to the biased galaxy formation idea the universe initially consisted of a roughly uniform cloud of cold dark matter and baryons. Within this cloud were small fluctuations, and as expansion continued "density peaks," or regions of excess density, were created, which, in turn, attracted other density peaks. These density peaks eventually led to bright galaxies and clusters.

But according to the theory this occurred only rarely. The computer program that simulates the universe is therefore written so that only large clumps of matter became galaxies. Galaxy formation from small clumps is suppressed, and therefore the program is "biased" in favor of the large galaxies. Galaxies therefore formed only in regions of peak density. Looking at such a universe would be like looking at an ocean. Here and there we see islands, but most of the mass is beneath the ocean. If this is the case in the universe it means that the large voids are filled with cold dark matter.

When the first computer simulations based on the biased galaxy idea were performed they gave distributions that agreed well with observations. "We can make direct pictures of models and the real universe and compare by eye and see whether they look the same," wrote White in a recent paper. "Overall there is no clear significant difference."

But not everyone is happy with the idea. Peebles isn't sure it will work in the long run. He is convinced that unless the cold dark matter has a speed near that of light, gravity will pull on all matter in the same way, and we will not get a good match with the large-scale structure of the universe.

The biased galaxy model is, however, supported by the

work of Gerhard Börner of the Max Planck Institute in Germany. He has been analyzing the distributions of galaxies in the sky using the latest data. "We found that the luminous galaxies tend to cluster," he said. "But fainter galaxies tend to be more uniformly distributed." A similar result was found by Neta Bahcall of the Space Telescope Science Institute and Raymond Soneira of the Institute for Advanced Study.

EXPLOSIVE GALAXY FORMATION AND COSMIC STRINGS

An entirely different approach to galaxy formation has been taken by several groups. They are convinced that galaxies, or at least most galaxies, did not originate from fluctuations. Jeremiah Ostriker of Princeton and Lennox Cowie of the University of Hawaii have suggested that most galaxies may have formed as a result of the explosive shock waves that are generated in young galaxies. Saito Ikeuchi of the University of Tokyo has made a similar proposal. In these models galaxies are created by other galaxies. When a galaxy is young it is well known that it is very active, with large numbers of supernovae occurring. According to the explosive formation theory these supernovae cause a shock wave that sweeps outward from the galaxy. Matter in the space around the galaxy is swept up as the shock wave passes it. As more and more matter collects it is compressed and eventually new galaxies form.

Supernovae occur in the new galaxies and new waves are sent out and new galaxies are formed. Eventually the shock waves from numerous galaxies coalesce into large expanding bubbles that compress the matter between them. The result is a picture similar to that seen by Huchra and Geller in their redshift survey. Huchra, in fact, says he favors this type of theory as an explanation of the bubbles he observes.

Despite its appeal, there are many difficulties. First, it is difficult to produce bubbles large enough to explain those seen by Huchra and Geller. Also, you might ask: If we need galaxies to produce other galaxies, where did the first one come from? On the basis of this it seems as if we still need the standard model—the theory that galaxies are produced from fluctuations. Perhaps a combination of the two ideas is best. And indeed this has been tried and appears to be reasonably successful.

Another approach to galaxy formation is based on cosmic strings. I talked about cosmic strings earlier, mentioning that they are thin, extremely dense strings that exist either in loops, or extend from one end of the universe to the other. They are so massive that a section an inch long would weigh as much as Earth.

There are two ways that strings could assist in galaxy formation. Both rely on the fact that they vibrate. Because of this motion they would produce gravitational waves that might compress clouds of gas and dust into galaxies, just as the shock wave did in the explosion theory. Better still, though, is an idea proposed by Jeremiah Ostriker, Edward Witten, and Chris Thompson of Princeton. First Witten showed that huge currents would pass along strings. Ostriker then picked up on the idea and showed that, because of the vibrations, cosmic strings could generate intense electromagnetic waves that would produce cavities around them, and in the process they would compress matter on the edge of the cavities.

The string idea has many possibilities but there is still a major problem: We have not yet detected a cosmic string. Indeed, most theories of structure still have problems. But there is little doubt that dark matter and the structure of the universe are linked. And, despite the problems, considerable progress has been made in the last few years.

CHAPTER 11

Gravitational Lenses and Dark Matter

Another potentially powerful tool in our search for dark matter is the gravitational lens, a gravitating object such as a galaxy or black hole that bends and focuses light much in the same way an optical lens does. Although the full potential of the gravitational lens has not yet been achieved, tremendous progress has been made, and it may one day provide the breakthrough we are looking for.

The bending of light by a gravitational field was predicted by Einstein in 1911. Shortly after publishing his special theory of relativity in 1905 (a theory that applied only to uniform motion in a straight line) Einstein began working on a theory that would generalize it to all types of motion. While working on this theory he discovered that light would be deflected by a gravitational field. According to his calculations the light ray from a star would be deflected by 0.83 second of arc (approximately the angle subtended by a dime at a distance of two miles) as it grazed the limb of the sun.

But how could such a deflection be observed? Stars are not visible when the sun is up. An eclipse seemed to fit the bill. Stars are visible during a total eclipse of the sun, and they can be photographed, and their positions determined. At Einstein's request the German astronomer Edwin Freunlich examined a number of plates that had been taken earlier, but he could find no evidence for deflection. What would be needed, he told Einstein, were measurements at an actual eclipse: photographs of the stars near the sun should be taken, and direct comparisons made to the known positions of these stars. He told him that an eclipse was scheduled to take place in Russia in 1914, and he offered to lead an expedition.

But Einstein was impatient; 1914 was three years away. He therefore wrote to George Hale at Mt. Wilson Observatory in California to see if the test could be made without an eclipse. Hale replied that it could not. So Freunlich set up an expedition, and early in the summer of 1914 he and his team set out for a spot in eastern Russia. But on August 1 Germany declared war on Russia and they were taken prisoner.

As strange as it might seem, it was fortunate that the test was not made at this time. Within a year Einstein had completed his general theory of relativity and was able to use it to recalculate the expected deflection. And lo and behold it turned out to be double his earlier estimate. If Freunlich had made the measurement and found double Einstein's prediction few would have taken his theory seriously.

The war was now on and there was little chance of another test for many years. The problem was compounded by the fact that almost no one outside of Germany had even heard of Einstein's new theory. Fortunately Holland remained neutral throughout the war, so there was a line of communication to the outside world. The Dutch astronomer Willem de Sitter had access to the German literature, and was also in communication with British scientists. And in 1916 he sent a copy of Einstein's theory to Arthur Eddington in England. It was a complicated maze of equations but Eddington quickly realized its importance. And, although the war was still on, he made preparations for a check of the theory at the next eclipse. Through his efforts two expeditions were set up, and in May 1919, one went to Brazil and the other to the west coast of Africa.

Despite poor weather Eddington and his group managed to get several photographs of the stars around the sun. And as predicted, they were deflected—not exactly by the amount Einstein calculated (1.7 seconds of arc), but when Eddington's results and the results from the other group were averaged, the prediction was within experimental error. Relations were still strained between Germany and England so it was several months before Einstein heard anything. Finally, though, on September 22, 1919 he received a telegram from H. A. Lorentz giving him the good news.

Then, perhaps because it was the first break in the strained relations between the two war-torn countries, the press pounced on it. Einstein's theory became an overnight sensation, and Einstein was soon famous throughout the world.

But what caused the deflection? Why did the beam bend? According to Einstein's theory the space around any massive object is curved. We cannot see this curvature, but a light ray passing through the space will follow the curvature, and therefore be deflected. If the light beam traversing this region comes from a star, its image will appear shifted slightly from its usual position.

But light rays passing a massive object are deflected on all sides of it. Is it possible that such an object could focus and perhaps magnify these rays as an optical lens does? Somebody asked Einstein this question in 1936, and he replied in a short article in *Science*. Einstein said that a gravitational lens was possible. "Of course, there's no chance of observing the phenomenon directly," he wrote. "The chance of two stars lining up is small, and even if they did we would not be able to observe it." He realized that exact alignment would be needed and that the distance between the objects was critical for optimum effect.

The paper passed almost unnoticed. One who did notice it, though, was Fritz Zwicky of Mt. Wilson. He didn't consider two stars in line with one another, but rather two galaxies, and found that although the probability of this happening was small, it was not zero. He then made calculations on the magnitude of the lens, and showed that if such a lens did exist he would be able to determine the mass of the foreground galaxy.

But the effect was beyond the capabilities of the telescopes of the day and therefore soon forgotten.

A CLOSER LOOK AT THE GRAVITATIONAL LENS

We've had a brief introduction to the gravitational lens, now let's look at it in a little more detail. We'll begin by considering the lensing of a quasar. In most cases the lensed object (the one being magnified) will be in the outer reaches of the universe. Since quasars are the most common objects there, it is reasonable that they will be involved. The foreground, or lensing, object can be any of several things, but let's assume that it's a black hole. If the effect is to be significant it's important that the lensing object be dense, and black holes fill this requirement. Furthermore, a black hole is so small it can be considered to be a point.

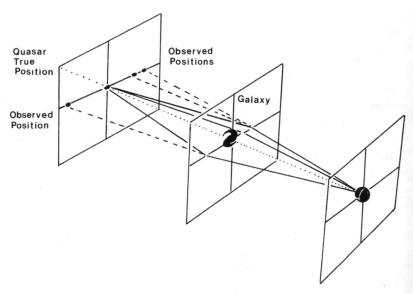

Schematic showing deflection of light around a galaxy—a gravitational lens.

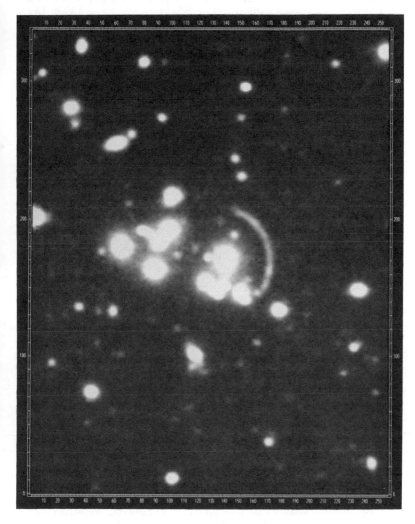

Portion of an Einstein ring. (Courtesy National Optical Astronomy Observatories.)

If there is an exact alignment between Earth, black hole, and quasar, the rays will diverge, bend around the black hole, then converge as they pass it, their paths meeting again at Earth. From Earth we would see the image of the quasar as a bright ring. In practice, though, it's much more likely that the alignment will be close, but not exact. What would we see in this case? If the alignment is off to one side, we would see two side-by-side images. Both would appear generally circular since they arise from a point source. If, on the other hand, the alignment was off in the vertical direction we would see two images, one above the other.

Black hole lenses are certainly the most exotic kind, but black holes are relatively rare in space; galaxies are much more common. What would happen if the deflector was a galaxy rather than a black hole? The major difficulty, of course, is that we would no longer be dealing with a point source, and as a result the image would be more complex. There may now be from three to five images, but again since they arise from a point source (the quasar) they will appear circular. We refer to this as a quasar–galaxy lens.

The last possibility we will consider is the galaxy–galaxy lens. We now have an extended object being lensed. If there was perfect alignment we would see what is called an Einstein ring—a circle of light. (An object of this type has recently been discovered by Jacqueline Hewitt of MIT.) The more likely case, however, is the imperfect alignment, in which case we would get two or more images of the galaxy. And in this case, since the images are extended, they would be kidney-shaped.

DISCOVERY OF THE FIRST GRAVITATIONAL LENS

Because it was so unlikely that two galaxies would line up there was little initial interest in gravitational lenses. But in the early 1960s the discovery of quasars rekindled interest in general relativity. It appeared that quasars might be massive objects undergoing gravitational collapse, or possibly black holes, and general relativity was needed to explain them. Interest in the theory soon led to an interest in gravitational lenses. Among the first to consider them was Sjur Refsdal, now of the University of Hamburg. He worked out the mathematics and found that gravitational lenses could become an important astronomical tool. He showed that they could be used to determine the velocity of expansion of the universe, and the average density of dark matter. The results were interesting, but without a lens to apply them to, few took an interest.

Then came an announcement that took the skeptics by surprise. Dennis Walsh of the Jodrell Bank radio-astronomy observatory, Robert Carswell of Cambridge University, and Ray Weymann of the University of Arizona published an article in the May 1979 issue of *Nature* announcing that they had detected what might be a gravitational lens. Using a telescope at Kitt Peak in Arizona they noticed that quasar 0957 +561 (the numbers refer to the coordinates of the quasar) appeared to consist of two components (now called A and B) separated by 6 seconds of arc in the sky. They had the same redshift, indicating a distance of approximately 6 billion light-years, and of particular importance, their optical spectra appeared to be virtually identical.

Were they just twins (they were, in fact, soon referred to as the "twin quasars") or were we seeing the first gravitational lens? Walsh, Carswell, and Weymann considered both possibilities. With the spectra being identical the case for a gravitational lens seemed strong. Nevertheless, there were problems. The lines in the spectra were of two types: dark lines (which are seen in stars and galaxies) and bright lines (usually seen only in gaseous nebulae). This was not surprising because both types are usually observed in quasars. But the redshift of the bright lines was different from the dark lines indicating that the quasar was surrounded by an expanding envelope of gas. And there was a problem relating this to a gravitational lens. If further progress was to be made, higher-quality spectra were needed.

Fortunately a powerful new telescope had just been completed. This telescope, called a multiple-mirror telescope, was of radically different design. It had been constructed on Mt. Hopkins in Arizona, and consisted of six 1.8-meter mirrors. With the help of computers and lasers the mirrors are synchronized so that they work as a single unit. They are equivalent to a single mirror 4.5 meters across.

On three evenings, from April 20 to 22, the MMT was trained on the twin quasar and new spectra were obtained. These were the first spectra taken by the new telescope and astronomers, although confident, were still unsure how well it would work. They were delighted, however, when the spectra turned out to be even better than expected. They were so good that astronomers could determine that the redshifts, and therefore the velocities, of the two objects were virtually identical.

The multiple-mirror telescope. (Courtesy MMT Observatory.)

A computer representation of the images formed in a gravitational lens. (Courtesy National Optical Astronomy Observatories.)

"I know they're twins, but are they paternal twins?"

Still, if the object was a gravitational lens, there had to be a "deflector" somewhere between us and the quasar. Searches were made in the region between them by Mark Adams and Tod Boroson of the University of Arizona using the 4-meter telescope at Kitt Peak. However, because the "seeing" was poor, the region was not well-resolved, and no deflector was found.

The twins then moved into the day sky and optical astronomers could no longer observe them. Radio astronomers, however, are not bothered by daylight, and they were able to take over observations. The first results came from the Very Large

The Very Large Array (VLA). (Courtesy National Radio Astronomy Observatory.)

Array (VLA) in Soccoro, New Mexico—a collection of twenty-seven 75-foot-diameter radio dishes laid out in a Y-shaped pattern covering an area about 25 miles across. The combination of the individual dishes is equal to one approximately 25 miles in diameter.

The VLA showed two images of roughly equal brightness corresponding to the optical ones. But detailed examination soon brought further problems. First of all, no deflector appeared to be visible between the images. And second, they were more complex than originally believed: besides images A and B there were three other faint images. As a result radio astronomers soon became convinced that the object could not be a gravitational lens. They were sure that it was a binary quasar. But the optical astronomers were sure it was a gravitational lens.

Finally, in November, the twins reappeared in the night sky

and optical astronomers went to work again. The major task now was to establish whether or not there was a deflector. If it was a gravitational lens there had to be one. Somehow it had just escaped detection. The first breakthrough came from Alan Stockton of the University of Hawaii. Using a telescope on Mauna Kea, Hawaii, he got a photograph under superb seeing conditions. It showed that the two images were not exactly identical; one appeared to have a faint fuzziness on its upper edge.

Then a group consisting of Peter Young, James Gunn, Jerome Kristian, Beverley Oke, and James Westphal hooked a powerful charge-coupled device (a silicon chip that converts light to an electric signal) to the 200-inch telescope at Mt. Palomar. Using it they obtained excellent images that also showed a fuzziness on the edge of one of them. But between the images there was nothing—no deflector. Careful examination finally convinced them that the fuzziness was actually a foreground galaxy superimposed on the image of the quasar. This was the deflector. They had expected it to be between the quasar images, but it wasn't. It was almost coincident with one of the images. This meant that things were more complex than previously assumed.

Theorists then got into the act and soon a new picture emerged. Calculations showed that there had to be three images; this meant that two of them were coincident and therefore unresolved. Careful examination of the images then showed that the deflector was an elliptical galaxy about halfway between us and the quasar. Furthermore, the elliptical was in a cluster. The dim images found by the VLA were some of the other galaxies in the cluster.

By mid-1980 astronomers were confident that the twin quasar was indeed a gravitational lens. The excitement of the discovery had barely started to die away when, much to everyone's surprise, another gravitational lens was discovered. Using the MMT, Ray Weymann and several colleagues discovered one with three images.

LENS SURVEYS AND DARK MATTER

Interest in gravitational lenses continued to mount after the discovery of the first lens. And it eventually led to several large-scale searches. In such a search astronomers begin by looking for quasars or galaxies that appear to have multiple images. (Radio telescopes are used in this part of the survey.) Once good candidates are found astronomers study them with optical telescopes.

The first search of this type was initiated by Edwin Turner of Princeton University and several collaborators. Turner received his undergraduate degree from MIT and his Ph.D. from the California Institute of Technology in 1975. Upon graduation he took a postdoctoral position at the Institute for Advanced Study in Princeton, and for a while was an assistant professor at Harvard University. From there he went to Princeton University.

"I got interested in lensing shortly after the first gravitational lens was discovered in 1979," he said. "The subject became fashionable . . . there was a lot of discussion about lenses so I decided to search for them." The switch into gravitational lenses was not a particularly large step for Turner. He had already worked in several closely associated areas: studies of quasar populations, evolution and dynamics of binary galaxies and galaxy clusters.

The idea for a large-scale survey came about as a result of discussions between Turner and Bernie Burke of MIT. "Burke had done some radio observations of the first gravitational lens," said Turner. "And I had done some theoretical work on it . . . calculations on how many we could expect to see. Bernie was already involved with several students in a big radio survey. He realized that his data could be used as the input to a lens survey. So it all got organized. . . we started about 1984."

Before its recent collapse, Burke and his group carried out the first step of the survey using the 300-foot radio telescope at Green Bank, West Virginia. They scanned the sky for

Edwin Turner. (Courtesy Robert Matthews.)

appropriate quasar candidates. These candidates are then looked at in more detail with the VLA. Radio "snapshots" are taken and inspected carefully. "We follow this up with optical images," said Turner. "If there are multiple optical images we take spectra. And if the spectra [of the images] are sufficiently similar we consider the objects to be a good candidate for a gravitational lens."

Turner went on to explain how gravitational waves would be useful in relation to the dark matter problem. "They will be helpful in two ways," he said. "First, by studying individual lens systems we can infer what the mass distribution in the system is and therefore can find out what the total mass distribution is—including the dark matter. Secondly, they will also be useful in a statistical way. If you look at the frequency of lenses [how often they occur] you get a statistical measure of the total amount of matter in the universe, including the dark matter."

The simplest way to visualize this technique is to think of the background quasars as seen through a large number of foreground objects—including dark objects. The more foreground objects there are, the more gravitational lenses there will be. Or conversely, the more gravitational lenses there are, the more dark matter there is.

Turner's work is also important in relation to the dark matter problem in that his results have enabled him to put limits on the abundance of specific forms of dark matter. For example, he and graduate student Jacqueline Hewitt established that there can be few black holes of galactic size.

Overall, about 20 gravitational lens candidates have now been found. Of these, 17 are quasar–galaxy lenses and 3 galaxy–galaxy lenses. "We have a total of 4200 radio maps and have looked at 200 candidates optically out of a total of about 400 we will be doing," he said. "So we're about half done, and we have found 6 or 7 lenses."

He went on to describe some of the problems of the research. "These things are enormously rare so the whole thing gives you the feeling of prospecting for gold. You spend an awful lot of time looking at sources that turn out to be uninteresting. We searched through a couple of thousand objects to find seven lenses." He paused for a few moments, then continued. "You go for long periods of time finding nothing, then suddenly you get something that's very exciting."

Turner also pointed out that although the average rate of discovery is only about two per year, the rate has been accelerating. "In the first few years only about one per year was

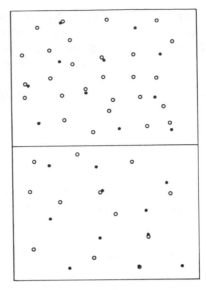

Gravitational lenses. The greater the density of objects, the greater the probability of a lensing object.

discovered. It's now up to three or four a year," he said. "And it's quite likely the rate will continue to increase. I wouldn't be surprised if it went from 20 to 100 in the next few years."

I asked him about the future of gravitational lensing. Will it eventually turn out to be a major tool of astronomy? "It's possible that if you came back in 30 years and looked at the field, astronomers might say it is one of the most important tools of cosmology. I also think it's possible that they might have lost interest in it entirely. They might say 'ya. . .that was a dead-end trail that never turned out to produce anything interesting.' Both are definite possibilities. The only way to find out is to try."

Another, slightly different type of survey is being conducted by Tony Tyson of Bell Labs. Unlike Turner who is interested primarily in quasars lensed by galaxies, Tyson is interested in galaxies lensed by other galaxies. Tyson did his undergraduate work at Stanford. Oddly enough, it was not in astronomy or physics, but in philosophy. He eventually became interested in gravitation, however, as a result of conversations with Leonard Schiff of Stanford, and decided to switch to physics. After obtaining his Ph.D. from the University of Wisconsin in 1967, he went to the University of Chicago for a one-year postdoctoral in astrophysics. From there he went to Bell Labs.

Between 1969 and 1984 he built four gravity wave detectors in an attempt to detect gravitational waves. But he was never able to detect them and finally decided to switch to astronomy.

Tyson is presently involved in two gravitational lens projects. The first one was initiated around 1980. He briefly described it to me. "We look at foreground galaxies, then go faint enough so that we can see lots of background galaxies, and if there is enough mass in the foreground galaxies, the background galaxies will be distorted. But you have to include a large number of foreground galaxies in order to get an effect. The distortion in a single foreground galaxy is extremely small. Galaxies just don't have enough mass to give a very large distortion to background galaxies. But if you have 10,000 foreground galaxies you can average the distortions of the background galaxies and come up with a significant effect." He went on to say that it would be another year or so before they had enough data to announce their results.

Of particular importance in relation to his work is the fact that the distortion gives you a measure of the amount of dark matter present. Tyson was, indeed, able to show that most galaxies do not have large dark matter halos. This is surprising in that, as we saw earlier, previous results indicate that galaxies do have extended halos.

How do his results fit in with earlier results? Tyson believes the apparent contradiction can be resolved if we assume that most galaxies are smaller than the Milky Way. He feels that large spirals have dark halos, but smaller galaxies do not. This means that "on the average" the halos will appear smaller than ours.

He admits, though, that his technique is far from foolproof, and as more observations are made things may change.

Tyson's second project involves looking for gravitational lenses in clusters of galaxies. "Clusters of galaxies have much larger clumps of dark matter in them," he said. "If you point your telescope at them you're much more likely to see lenses. In fact, literally every compact cluster we have looked at so far showed some evidence of systematic distortion of the background galaxies."

CHAPTER 12

A Consensus: To Be or Not To Be

Now that we've surveyed the evidence for dark matter it's time to summarize things and come to a conclusion. The first thing we can say is that many problems remain, and we're still far from a solution. In fact, we're still not sure there is a dark matter problem. And if there is we're not sure what the dark matter is. True, we have a number of candidates, but each of them presents problems. Furthermore, we don't know how the dark matter is distributed—whether it is uniform throughout the universe, or in clumps.

Let's begin with the question: Is there really a dark matter problem? Virginia Trimble answers this with: "Not necessarily, though there are clearly a number of astrophysical problems [e.g., evidence for halos around galaxies] to which different kinds of dark matter are among the best possible solutions." The astrophysical problems she is referring to, most astronomers would agree, do indicate that there is dark matter associated with galaxies and clusters. In contrast, the dark matter, or perhaps I should say the missing mass, of the overall universe may or may not exist. It is quite possible that the universe is open, in which case there would be no missing mass.

Furthermore, not everyone believes that there is dark matter even in galaxies and clusters. Mordehai Milgrom of the Weizmann Institute in Israel has suggested that there appears to be dark matter in galaxies and clusters because we are looking at the problem in the wrong way. He believes that the laws we are using to make our calculations may be wrong.

We know, for example, that Newton's laws of motion are valid for everyday motions involving cars, billiard balls, and even rockets. Yet if we try to apply them to the microcosm (the world of atoms and elementary particles) they don't work. They give us the wrong answer. On the other hand, when we deal with large structures such as galaxies, we are dealing with distances just as far removed from the everyday distances of our experiences as atoms are. Is it possible that Newton's laws break down here also?

There is strong evidence that his laws of motion are valid, but his law of gravity is worth looking at a little closer. Milgrom believes that when gravity becomes exceedingly weak—millions of times smaller than the gravitational pull between Earth and the sun—Newton's law of gravity may no longer be valid. He has shown that if the force falls off less rapidly than Newton prescribed he can explain many of the problems usually associated with dark matter.

Actually, a slight modification of Newton's law is nothing new. Einstein modified it in his general theory of relativity. He showed us that it is not valid when the gravitational field is exceedingly strong. Perhaps it's also violated when the field is exceedingly weak. In the early 1980s Milgrom wrote several papers showing that this could explain many of the problems that we associate with dark matter. And in December 1986 he reported his results at a conference on dark matter in Jerusalem. He began his talk with the statement, "I want to present you a solution to the mass discrepancy problem that is utterly different from the conventional explanation." He then briefly outlined his plan.

"What is the motivation for such a route?" he asked. Answering his own question he said, "The conventional solution—the dark matter hypothesis—leaves much to be desired. [It] is completely arbitrary in that it invokes the existence of dark matter in just the correct amount and spatial distribution needed to explain the mass discrepancy in each and every case." He then reminded the audience that "not a trace of dark matter has been detected directly." "In short," he said, "the dark matter hypothesis is very elaborate and yet quite useless."

With his modification Milgrom has been able to solve many of the problems associated with dark matter: the mass discrepancy in galaxies, clusters, and superclusters. And the flat rotation curves of Rubin and others.

This is such a reasonable and easy way out of the difficulty it seems that most scientists would eagerly accept it. But this hasn't happened. Most scientists are, in fact, quite skeptical of it; some consider it downright heresy.

One of the major objections to any change in Newton's law has to do with one of its most beautiful properties. Consider a hollow spherical shell of uniform matter. According to Newton's theory it will exert no gravitational pull on anything inside it. The forces are the same in all directions, and will therefore cancel, giving a net force of zero. This means that if we consider the universe to be made up of a series of concentric shells, like the skins of an onion, the outer shells will have no effect on the inner ones. Because of this we can select a small isolated section of the universe and study it as a representation of the overall universe. We need not know how the entire universe works: whatever happens in our small section will happen everywhere. But when we start tampering with Newton's law this is lost. Milgrom has shown, however, that this argument doesn't apply to his theory. His changes are so small that they avoid this difficulty.

Nevertheless, there are problems. One is the difficulty of testing a small modification of Newton's law over galactic distances. We can easily test Newton's law in the solar system because the effects are large. Even in the case of two stars moving around one another we do not have a serious problem. But when we attempt to test the theory for two galaxies orbiting one another . . . well, that's a different matter. Even over thousands of years galaxies barely move; it would therefore be extremely difficult to notice the effects of a slight variation in Newton's law.

In order to get a better perspective on Milgrom's theory I talked to two astronomers who are quite familiar with it. The first was John Bahcall of the Institute for Advanced Study. Milgrom formulated his theory while at the Institute. "Milgrom's work is extraordinary and beautiful because it makes a connection between things that don't otherwise appear to be connected," he said. "What is really quite striking is that most of us here at Princeton thought the idea would be shot down in six months. But somehow it has been much more robust than we anticipated. Its main virtue is that it has shown us how little we really know about gravity. The fact that it hasn't been shot down somehow indicates that there's something we don't understand." He paused, then finished a little reluctantly with: "My guess, though, is that it must be wrong."

David Schramm of the University of Chicago is also convinced that it is wrong. "I think it's a suggestion worth exploring," he said. "But it has a serious drawback: it's really a local

John Bahcall.

solution. When you're talking about cosmology you've got to encompass the entire universe. Milgrom's model breaks down at really large distances, so you know it can't be the true cosmological model. If you try to extrapolate it to distances known to exist in the universe you end up with peculiar behavior."

Despite the controversy, several scientists have taken it upon themselves to try to disprove the theory, and so far they haven't been able to. So, while it appears unlikely it will ever replace Newton's law it has generated considerable interest.

This takes us back to square one. If we can't explain the problems without invoking dark matter, then dark matter must exist . . . or so it seems. Let's turn, then, to this case. In reality the problem consists of three parts. The questions we asked at the beginning of the chapter should be answered for each of:

1. The dark matter associated with the disk and halo of our galaxy (and other galaxies)
2. The dark matter associated with clusters and superclusters
3. The missing mass (assuming it is missing) of the overall universe

Let's briefly review each of these problems beginning with our galaxy. As we saw earlier, Jan Oort of Holland showed in 1932 that there was about 50 percent more mass in the Milky Way than could be accounted for. This problem has still not been solved. It has, in fact, been substantiated by recent work. John Bahcall redid the experiment in the mid-1980s with more modern equipment and techniques. "What I did differently," he said, "was use a large computer to do the calculation. I also had a much better model of our galaxy, and I took into account the dark halo around it." With the computer he was able to look at many aspects of the problem that were beyond the capability of Oort.

I asked him how his results differed from Oort's. "Jan concluded that there was roughly comparable amounts of dark matter and luminous matter in the vicinity of the sun. My conclusion is consistent with this, but I got a range of values. I found that the dark matter can be between 0.5 and 1.5 times the amount of luminous matter."

What does he think the dark matter is? "All available observations are consistent with the conservative interpretation that the missing dark matter is mostly brown dwarfs," he said. But he admits that until there is proof, we cannot be sure.

In addition to this problem we saw earlier that Vera Rubin and her colleagues showed that the velocity curves of galaxies indicate that galaxies have huge halos. What are these halos composed of? We still don't know, but we saw in the last chapter that there have been some interesting developments. Using gravitational lenses Tyson has shown that all galaxies do not appear to have gigantic halos; it may be that only spirals such as ours do. Furthermore, Stephen Kent of the Harvard–Smithsonian Center for Astrophysics has recently shown more directly that many galaxies have halos much smaller than previously assumed. Using a CCD he studied the rotational curves of 37 spirals and found that in most cases the curve dropped off just past the outer limit of visible material, indicating a relatively small halo.

Continuing on to clusters of galaxies we find that evidence still suggests that the dark matter in small clusters exceeds the luminous matter by a factor of 10 to 30. And when we go to large clusters this number climbs to several hundred. This means that even if we agree that each of the galaxies has a considerable amount of dark matter throughout it and in the halo around it, we are still far short of the amount of dark matter that appears to exist in large clusters. These clusters are too massive by a factor of 10 or more, which means that they contain about 99 percent dark matter.

Finally, when we look at the overall universe, things are even worse. Inflation theory tells us that it should be just closed, which means that its average density is equal to its critical density. If so, the amount of dark matter in the universe may be a thousand times the amount of luminous matter.

THE CANDIDATES

Let's turn, then, to the candidates themselves. We've had a close look at all of them and should be in a good position now to make a final judgment. As we saw earlier the favorite candidate among astronomers is ordinary matter—protons and neutrons—in the form of brown dwarfs. But even here there's a problem: according to nucleosynthesis we can't close the universe with ordinary particles (baryons). We saw earlier, though, that there are ways around this. First of all, it is possible that the universe is closed with particles other than baryons— the so-called exotic particles. Second, there's inhomogeneous nucleosynthesis. The usual nucleosynthesis calculations are performed under the assumption that the universe was a smooth, uniform gas. But if this wasn't the case—i.e., if the universe was "lumpy" at the time of nucleosynthesis—then it may be possible to close the universe with baryons. Several astronomers are presently trying to do this. But recently they have run into serious problems. Schramm summarized the latest results for me. "Initially it looked as if we might get around the problems," he said. "But the latest work indicates that we might not. We have always had problems predicting the proper amount of lithium, but now the more detailed calculations show that helium is overproduced. We thought we might get around the lithium problem with some sort of additional process that would deplete it . . . some stellar process. But with the helium problem it is clear you can't do it. Work has been done by several groups, one in California, one in Texas, and one in Japan. And they all come to the same conclusion."

Schramm is therefore skeptical that such theories will succeed. Bahcall, on the other hand, is not convinced the nucleosynthesis limit will stand. "Some of us regard the nucleosynthesis argument like church attendance: Obligatory in good society, but not to be taken seriously in everyday life," he said.

If inhomogeneous nucleosynthesis fails, and this now appears likely, we have to look to the exotic particles. They do, indeed, get around most of the difficulties, but they themselves are a problem: So far none of them have been observed. In other words, they exist only on paper. Furthermore, some of them are predicted by grand unified theories (GUTs), and GUTs now seem to be in trouble.

In Chapter 10 we saw that when exotic particles (in particular, cold dark matter) are used in models of the early development of the universe, they give excellent agreement with observation. Does this mean we can assume the dark matter is exotic particles? At this stage we certainly can't, since we don't know for sure they exist. Furthermore, there is another problem. Earlier I mentioned that there are, in reality, three different dark matter problems. The best solution would, of course, be to have one type of dark matter solve all three problems. But exotic particles don't. It is possible that they are the dark matter of galactic halos, the dark matter in clusters, and the missing mass of the universe, but observations of the gamma ray intensity within our galaxy indicate that they can't explain the dark matter within it (the gamma ray intensity is too low).

The same sort of problem prevails in the case of all candidates. Brown dwarfs, for example, can explain the dark matter of our galaxy and its halo, but they can't explain the missing mass of the universe. Massive neutrinos can explain the halos around galaxies, and even clusters, but if we assume fluctuations gave rise to the galaxies, they can't form galaxies properly. Furthermore, they can't explain the dark matter in the disk of our galaxy; they move too fast. And so it goes.

It is, of course, possible, that there are two types of dark matter, or even more. But astronomers would prefer the simplest solution, and that is a single type of dark matter that solves all problems. The likelihood of this, however, does not appear great at the present time.

THE SEARCH FOR EXOTIC PARTICLES

Exotic particles have become such a serious candidate within recent years that several searches for them have been initiated. Searches for photinos, for example, have recently been made at several large particle accelerators, called colliders. As their name suggests, particles in such an accelerator collide head-on. This is possible because all particles have what are called antiparticles associated with them. The antiparticle to a given particle has the same mass, but (in most cases) it has an opposite charge. Corresponding to the electron, for example, there is a positively charged antiparticle called the positron. When the electron and the positron are brought together they annihilate one another releasing energy in the form of photons.

In a collider, particles and their antiparticles are set in motion inside a large ring. They move in opposite directions because of their opposite charge. At the appropriate time adjustments are made so that the two beams collide.

If the participants of, say, an electron–positron collision have enough energy when they collide they can produce a pair of superparticles: a selectron and an antiselectron. This pair will rapidly decay to lighter particles and superparticles. Some of the superparticles will be photinos, and because they are the lightest of them they cannot decay. Photinos should therefore be detectable.

Colliding beam reactions of this type have been performed at Cornell, Stanford, and in Germany, but no evidence for the photino has been found. Similar searches using protons and antiprotons have been made at CERN. In this case some evidence was found, but most physicists now believe it was likely due to some known particle such as the neutrino.

Searches for the axion are also in the planning stage. Pierre Sikivie of the University of Florida has shown that they might be detected in microwave cavities. According to his calculations, oscillating axions might produce electromagnetic radiation that could be detected. Frank Wilczek and John Moody of the University of California at Santa Barbara, Don Morris of Lawrence Berkeley Labs, and Lawrence Kraus of Yale have made refinements to Sikivie's scheme, and have suggested a detection scheme of their own. So far, though, no axions have been found.

In another experiment Blas Cabrera and several colleagues at Stanford University are attempting to detect dark matter directly. They have shown that if there is dark matter throughout our galaxy, we will be moving through it, and will therefore experience a "dark matter wind." They are building an apparatus that may enable them to measure this wind. The heart of their apparatus consists of a large block of silicon with detectors attached on all sides. The entire apparatus is cooled to liquid helium temperatures.

They believe that as this "wind" of dark particles sweeps across the surface of Earth some of the particles will occasionally strike the atoms of silicon in their apparatus, causing them to vibrate. But each atom of silicon is attached to many other atoms around it, and if one begins to vibrate it will cause its neighbors to vibrate. If this disturbance continues all the way to the surface it will be picked up by one of the detectors. It sounds like a simple enough technique but in practice it is much more difficult to do than it looks. The major problem is that many different types of particles will pass through the apparatus and cause atoms to vibrate—neutrinos from the sun, for example. They will somehow have to be distinguished from those caused by dark matter. Cabrera believes he has this problem solved, however, and if so he may soon be detecting dark matter.

Joseph Silk of the University of California is also involved in several dark matter detection experiments. "My involvement is mostly theoretical," he told me. "The experiments are very

Joseph Silk.

difficult and they're really just beginning. In the first type we'll be looking for dark matter in the halo. Dark matter is not really completely dark; every now and then dark matter particles annihilate with themselves. And you can calculate what the annihilation rate is." He explained that the annihilation rate was high when the dark matter was first formed in the early universe, but that it is extremely low today. "Nevertheless," he said, "you can still see some effects of it because when a dark matter pair does annihilate a lot of energy is released. And it should show up in the form of gamma rays or cosmic rays."

He described two types of experiments to me. In the first, astronomers will search for gamma rays that result from the annihilation of dark matter. This experiment will require a large detector, and it will have to be put in space. A serious problem with it is that gamma rays are produced in other processes, and they will have to be distinguished from the ones produced by dark matter.

In the second type of experiment, astronomers will search for antiprotons. They are also produced in the annihilation of dark matter. "Antiprotons are normally rare in cosmic rays," said Silk. "So if you find a large flux of them you should be able to associate them with dark matter. They can be distinguished from the antiprotons produced by cosmic rays in that they are much less energetic."

Another interesting possibility is that dark matter interacts with our sun. "Dark matter particles that run into our sun will get stuck in it, and sink to the middle," Silk said. "This gives it an extra source of energy, which is very weak compared to what is produced by thermonuclear reactions, but it will have a slight effect. One way to detect it would be to look at the neutrinos that are produced by the sun. Nuclear reaction neutrinos have already been found . . . they come right out of the core. The neutrinos that result from dark matter will be much more energetic and we should therefore be able to distinguish them. There are a number of experiments planned for doing this." He concluded by saying that any of the above experiments will take at least ten years to complete, so the problem will not be solved for a while.

OPINIONS FROM THE EXPERTS

In an effort to get a final overview of the dark matter problem I talked at length to David Schramm and Joseph Silk. They are acknowledged experts, and have been working in the area for several years.

Schramm was on the team that showed in the 1970s that nucleosynthesis would not allow the universe to be closed by

David Schramm. (Courtesy Patricia Evans.)

baryons. "The dark matter problem is certainly one of the forefront problems of cosmology today," he said. "I think the key that made it such an important area is that it's not just a theoretical one. Cosmology is now interfacing with experiment and observation . . . which is quite different from what it was 30 years ago. I think there are enough experiments under way right now that we will, within a decade, either find the dark matter, or find that cold dark matter doesn't exist. Also, we've got accelerators looking for different kinds of matter. If these experiments come up with something, they might give a solution to the problem. If they don't find anything, though, it's going to make it more frustrating." He paused, then continued, "Yes. . . it's a long-range problem and the real resolution won't come until we find the stuff in experiments or in accelerators . . . assuming it's not baryons."

He then went on to talk about the difference between the various dark matter problems, pointing out that the dark matter of halos need not be exotic particles. "We showed that what is baryonic is in pretty good agreement with what you see if you include dark halos," he said. "So the dark halos could very well be low-mass stars and that would still be consistent with nucleosynthesis. Some people talk as if the halos have to be made out of weird stuff [exotic particles] and yet the data doesn't require it. It always seems that when you start talking about the weird stuff someone immediately says, 'Let's make the missing mass of the universe and the dark matter of the halo the same thing.' But you're not really forced into this at all."

The conversation then turned to the making of galaxies and structure in the universe. "Different scenarios tend to favor different kinds of dark matter," he said. "If you have a scenario that has small random fluctuations appearing at the end of inflation, the favorite candidate is cold dark matter [axions, photinos, and so on]. It has some nice features and is perhaps the simplest model, but it also has some difficulties. If, instead of quantum fluctuations, you have cosmic strings giving the galaxies, the mechanism is quite different. It turns out that instead of cold dark matter being the preferred model, the hot dark matter [massive neutrinos] model is better. Another model that has been popular lately is one in which you have an explosion of cosmic strings. In this case galaxies form where the matter is compressed. This will work with either hot dark matter or cold dark matter."

I asked him about the arguments from inflation theory that imply that the universe has a lot of missing mass. "While the details of inflation aren't firmed up yet, and it still requires some fine-tuning, the basic underlying idea is very good," he said. "I think you need something like that. It provides a way of setting the initial conditions of the universe. So, while we might not know the details yet, I believe we've got basically the correct picture."

Silk, on the other hand, still has some reservations. Regarding inflation he said, "Some people believe it and some don't. There are theories that are incompatible with it that some people believe . . . like cosmic strings. The cósmic string theory is quite nice for explaining large-scale structures and if you believe it you don't have room left for inflation. They're incompatible."

Regarding large-scale structure he said, "Dark matter is very important for making large-scale structure because it's such a large component of the overall gravity of the universe. In fact, it's very difficult to make large-scale structure if you don't have dark matter." He paused, then continued, "There are several ways you can make this structure. Cosmic strings are popular right now. But you have to have some form of matter that will accrete onto them to make galaxies. It could be ordinary, or baryonic, matter in which case the universe would likely be open. It doesn't need to be closed if you're using strings. Or you could have baryons and cold dark matter, or baryons and hot dark matter along with cosmic strings."

Which is best? I asked. "If you go through all the models people have come up with you find it's probably easiest to make objects that look like galaxies if you have a combination of strings along with baryons and hot dark matter. You need the dark matter to explain the halos of galaxies."

But is the dark matter in the halos and clusters the same as the dark matter of the universe? "There's not enough dark matter in clusters to close the universe," Silk said. "It's only about one-tenth the amount needed. Of course we don't know for sure the universe is closed. But if there turns out to be ten times as much in clusters as we now think, then obviously the two types of dark matter would be the same. And we can't rule that out. But it could equally well be something quite different."

A NEW REVOLUTION?

The dark matter mystery is, without a doubt, one of the central problems of cosmology today. And the reason is quite clear: Until it is solved—until we know exactly how much dark matter there is—we will not know the fate of the universe. And the fate of the universe is certainly one of the foremost problems of cosmology. Furthermore, this isn't the only problem. Until we know what form the dark matter takes we can't thoroughly understand the creation of galaxies, their structure, and even the overall structure of the universe.

With so many unsolved problems something is clearly wrong with some of the ideas we now have. This makes us ask: Are we on the brink of a significant breakthrough as we were in the early 1900s just before Einstein published his theory of relativity, or in the 1920s before the birth of quantum theory? Will a new theory come forth, or a new observational discovery be made, that will bring all the pieces together giving us a beautiful and simple solution to our problems?

There is, indeed, an analogy. In the early part of the century something was drastically wrong with our model of the atom. If electrons whirled in orbit around a nucleus, as Rutherford's model indicated, they had to radiate away their energy. A basic law of electromagnetic theory tells us that any accelerating charged particle radiates, and it is easy to show that, according to the theory, atoms could last no longer than a few millionths of a second. Yet there are atoms all around us—and they had been around for billions of years. Something was wrong. And indeed, Niels Bohr soon showed us we were looking at things in the wrong way. A new principle had to be applied.

Today we have galaxies that don't rotate as current theory says they should. We have clusters of galaxies with too much mass. We have models of the formation of the universe that are not satisfactory. And the overall universe seems to be missing over 99 percent of its mass.

Is cosmology in the same position today that physics was in about 1920? It's hard to say. We can say, though, that it's unlikely that the solutions will come with a sudden all-encompassing discovery. From past experience we know this just isn't the way things work. Even if an important breakthrough comes it takes time to assimilate and accept it. It's more likely that the problems will be whittled away—perhaps one at a time. Important breakthroughs will come—some of them no doubt will be important observational discoveries, others theoretical breakthroughs. But it will take time to see their full implications.

While I'm on the subject of new theories and breakthroughs let me say a few things about a serious problem. If progress is to be made—particularly on the theoretical side—crazy new ideas will be needed. And such ideas usually come from young theorists. I say "young" because the most creative years of a theorist are usually the first ten after he receives his Ph.D. Einstein, Bohr, Heisenberg, and Pauli were all most productive before the age of 35. This means that a theorist must take advantage of this time; he must establish a reputation with a few important and innovative papers if he is to get a good position at a prestigious university. This means that he must develop new ideas and yet at the same time they can't be too crazy or he may be labeled as a lunatic, or at least not taken seriously. This has happened in a number of cases. We talked about the coolness toward Milgrom's idea earlier; Arp also ran into the same problem when he suggested that all redshifts may not indicate recession. That's not to say, of course, that every crazy idea has to be considered seriously. There are a lot of crazy ideas put forward each year that aren't worth the paper they're written on. So we have to be able to sift out the "good" crazy ideas from the bad ones. And that isn't easy.

What this all means is that the young theorist must be innovative and yet must restrain himself from straying too far from accepted ideas. It's a kind of a catch-22: you can't progress without new ideas, but if you go a little too far with your crazy new ideas you're an outcast.

Anyway, the important point is: new ideas are needed. And new observational discoveries are needed. Without them we will soon stagnate. How long will it take to solve the dark matter mystery? Will it drag on for 20 or 30 years or more? We still don't know. Will it bring a revolution in cosmology? John Bahcall answered this with: "I don't know. On some days I think we really don't understand a large part of the universe. On alternate days I think the solution will be something very conventional like small stars. I just don't know."

CHAPTER 13

Fate of the Open Universe

One of the major reasons for studying the dark matter mystery is that the fate of the universe depends on how much dark matter there is in it. Since we now have some understanding of the problems involved, we will direct our attention throughout the rest of the book to the fate of the universe. In an earlier chapter we saw that its fate depends on its average density. If it is under a certain critical density it will expand forever; if it is over this density it will eventually stop expanding and collapse back on itself. It is important, therefore, to determine the average density of the universe accurately. And this, in turn, depends on how much dark matter there is. Is there, in fact, enough dark matter to close it (stop its expansion)? As we have just seen, we're still not certain, but we do know that it is close, and because of this I will discuss both the open and closed universes. This chapter will consider the open universe.

One of the things that will no doubt surprise you is the exceedingly long times we will be talking about. It's difficult enough trying to imagine a time span of, say, a million years. What will Earth look like in a million years? What changes will have occurred? Will there even be a civilization here? It's hard to say; if we're honest, though, we have to admit that we can only guess. After all, we've only been around for a few thousand years and significant changes have occurred even in the last 100 years. Yet, as long as a million years seems, it is insignificantly short compared to the times we will be talking about. Even billions of years will seem short in the latter part of the discussion.

Why are we interested in such long times? The major reason, I suppose, is that it's human nature to want to know what will eventually happen to Earth (and the universe), even though civilization will be gone long before it comes to pass.

You might also ask how we know our ideas are correct. At this point I'll have to admit we're not sure all of them are. Science (particularly the frontiers of science) is always changing, so in 100, or even 50, years it's quite possible we will have different ideas. Some of what I will say is indeed speculative, or perhaps I should say, unproven, but as much as possible everything is based on the laws of physics as we presently know them. This, of course, brings up the questions: Are we certain we know all the laws? And are the laws independent of time—will they remain the same for billions of years? Again, we're not sure, but until we know better we have to assume they are valid. If we find out later that modifications are needed we can easily make them, just as we have in the past.

A major modification of our ideas about the future of the universe did, in fact, occur early in this century. Before the discovery of the expansion of the universe scientists believed that the universe was doomed to what they called a "heat death"— an equilibrium situation in which everything got colder and colder, eventually reaching 0 K. We now know that things aren't this simple. When the force of gravity and the expansion of the universe are taken into consideration we find that equilibrium is never reached. The constituents of the universe eventually get so far apart they no longer interact.

What is perhaps most surprising about the future of the universe is that so few scientific papers have been written about it. Yet thousands of papers have been written about its creation. Two people took up the challenge of filling this void in the mid-1970s: Jamal Islam of Cambridge University, now at the University of Chittagong in Bangladesh, and Freeman Dyson of the Institute for Advanced Study in Princeton. Islam broke the ice with a paper in 1979 that outlined many of the events that will occur in the distant future. Dyson picked up on his work and extended it.

Born in England in 1923, Dyson became interested in relativity and astronomy at an early age. He read popular books on Einstein's theory while still in his early teens. Excited about what he read he decided to get enough background to understand the theory in depth. But calculus and differential equations were not taught in secondary school so he decided to order a book and learn about them on his own. Spotting what he thought was an appropriate book in a catalogue one day, he sent for it. Much to his delight it arrived just before he was scheduled to go on vacation. The book soon became his prize possession. Each day he would rise at 6:00 A.M. and begin studying it, working his way through the exercises page by page. It was seldom closed before 10:00 P.M. He spent so much time studying it his mother began to worry. "You'll ruin your health and burn out your brains," she told him. But not until the vacation was

Freeman Dyson.

almost over, and he had worked his way through the entire book, did he put it away. Looking back on it later he said, "It was the best vacation I ever had."

In 1947 Dyson came to Cornell to work with Hans Bethe. Richard Feynman was also there and they soon became close friends. One of Dyson's most important papers, in fact, came about as a result of this friendship. Feynman and, independently, Julian Schwinger of Columbia had just solved one of the major problems of physics. They had made the theory of electrons and photons (quantum electrodynamics) into a viable theory. But strangely they had used what appeared to be entirely different approaches. Dyson familiarized himself with both approaches and showed that they were equivalent.

Dyson eventually became interested in extraterrestrial life and the fate of life over long periods of time in the universe. This led him to take a close look at the equations governing the future of the universe, and in 1979 he published a paper titled "Time Without End: Physics and Biology in an Open Universe." In it he divided the future into several stages, describing the events of each stage in detail. I will follow his approach.

STAGE 1: THE DEATH OF STARS

The first event in what might be called the death of the universe is the demise of the stars. A star forms when a large cloud of hydrogen and helium (contaminated with about 1 percent of all other elements) contracts under the force of gravity. As it contracts its core heats until it reaches a temperature of about 15 million degrees. At this point the hydrogen is ignited, and the ball of gas becomes a star. Radiation rushes out from its core creating a force that, along with its gas pressure, balances the inward gravitational pull. The star stops contracting, and for millions, or perhaps billions, of years it remains essentially unchanged.

Before continuing I should clarify something. I referred to

an "ignition" above, which seems to imply that the hydrogen begins to burn. And, indeed, in a sense, it does. More exactly, though, what is "ignited" is a sequence of nuclear reactions that convert hydrogen into helium. We sometimes refer to the helium as the "ash" of the burning hydrogen.

Now back to our story. The helium that is produced is heavier than the hydrogen and therefore falls to the center of the star. Gradually, as more and more hydrogen is burned a core of helium builds up, and as it grows the pressure at its center increases. This, in turn, increases its temperature until finally the ignition temperature of helium (approximately 100 million degrees) is reached, and the helium begins to burn.

In a star about the mass of our sun the ignition of helium is accompanied by an explosion. But it is not a powerful enough explosion to blow the star apart; only the core is disrupted. And over a few hundred years the helium falls back to the center and begins to burn peacefully. In the same way hydrogen burning begins again in a shell around the helium core.

But the burning helium also leaves "ash" just as the burning hydrogen did. In this case it is a mixture of carbon and oxygen, and again it accumulates at the center of the star. And if the star is massive enough this new core will eventually begin to burn.

Will all stars trigger their core of helium (or carbon–oxygen, if they have one)? No, if the star is small its core will never get hot enough to burn helium. Our sun, however, will; it is currently burning hydrogen with helium accumulating in the core. It will eventually burn the helium, but it will burn nothing further. If a star is massive enough, though, it can burn elements all the way up to iron. This is, in fact, how the elements of the universe are produced.

But stars only have so much fuel. Massive stars have the most, but they use it up at such a fast rate they live only a few million years. Tiny red stars, on the other hand, have little fuel, but they burn it so slowly that they can last as long as 50 billion years. This means that 50 billion years from now there will be no

giant stars left, only small red dwarfs, and our galaxy will have
grown dim. The stars in it will, one by one, flicker and go out
until in about 100 billion years it will no longer be visible.

But what happens to the stars when they die? As I
mentioned earlier, their fate depends on their mass. Relatively
small stars like our sun collapse slowly over millions of years
after their thermonuclear furnaces go out. They end as white
dwarfs—objects not much larger than Earth. But there is a
limiting mass for white dwarfs. If a star has a mass greater than
1 ¼ solar masses it will not end as a white dwarf. It will explode
as a supernova, and its core will be crushed to a neutron star—
an object considerably smaller than a white dwarf. On the
average, neutron stars are only a few miles across. And finally,
the most massive stars will collapse to black holes.

This means that our galaxy will eventually consist of white
dwarfs, neutron stars, black holes, and a few smaller objects
such as planets and asteroids. I should also mention that while
the stars are dying, galaxies will be moving away from us until
finally, if there were anyone here to observe them (and, of

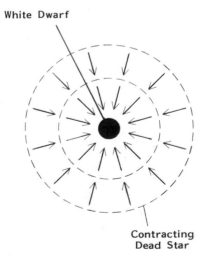

Contraction of a star to a white dwarf.

course, there won't be), we would see no galaxies, even in the largest telescopes. In fact, we wouldn't even see stars—they will no longer be luminous. All this will have come to pass by the year 10^{14}.

STAGE 2: STARS LOSE PLANETS

While the above events are going on, other changes will also be occurring. One of them is the loss of planets by stars. We know that our sun has a system of nine planets. And although our system might be larger than most, it seems likely that most stars have planets. We can, in fact, indirectly detect dark objects around several of our nearby stars. Barnard's star, our second nearest neighbor, for example, appears to have two dark objects associated with it. Indeed, if our sun was the only star with a system of planets it would be special, and astronomers don't believe that it is. They have, in fact, formulated what is called the principle of mediocrity stating that we are not special in any way. Assuming this is the case, it seems reasonable that most other stars have planets.

Starting with this assumption Dyson calculated how long a typical star would keep its planets. Without delving into the details of his calculation it is easy to show it will be extremely long. The stars in our galaxy orbit harmoniously; each has a particular speed that depends on its distance from the center, much in the same way the planets in our solar system do. And, on the average, they are separated by about 5 light-years. In the 15 billion years or so that our galaxy has existed, only a few stellar collisions have occurred in its outer regions. This means that few planets have been ejected from their system. But in, say, a million-billion (10^{15}) years the number of collisions will be quite significant. In fact, since you don't actually need a direct collision to knock a star's planets out of orbit (a close encounter will do) we might expect a large number of planets to be gone. And, indeed, they will be; according to Dyson such encounters

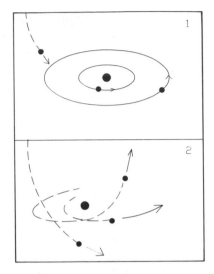

The loss of planets due to collisions.

will clean most stars of their planets in about 10^{16} years. By this time, of course, all stars will be dead.

STAGE 3: GALAXIES LOSE STARS

Most of the "collisions" that occur between stars in a galaxy will only be close encounters. And although they are sufficient to free planets, they will not disturb the star itself. Occasionally, though, an encounter will be close enough to literally knock the star out of orbit, and in many cases, completely out of the galaxy. Such collisions will, of course, be much rarer than those that knock planets out. But again, over a very long time they will be important.

To an outsider it will appear as if our galaxy is slowly evaporating. Over billions of years it will grow smaller as its outer stars are ejected. But the escaping stars take energy with them so our galaxy will gradually lose energy, and as a result the remaining stars will begin to fall to the center. This, in turn, will

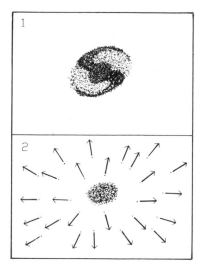

Evaporation of a galaxy.

accelerate the decrease in size. The remaining objects will get closer and closer together as they fall. Many of them will be black holes, which will pull in the dead stars around them. And as these black holes come together they will coalesce, generating even larger black holes.

But the largest black hole will be at the center. For years it has been assumed that radio galaxies have supermassive black holes at their centers that are responsible for their radio emission. Many astronomers are convinced, in fact, that all galaxies have massive black holes at their center. If so, many of the smaller black holes, neutron stars, and white dwarfs around it will be dragged into it. Eventually it will have a mass of about 10^{10} solar masses.

Dyson has shown that by the year 100 billion-billion (10^{20}) 10 percent of our galaxy's mass will be in the form of a supermassive black hole. The remaining mass will be in smaller black holes, black dwarfs, neutron stars, and so on that surround the giant black hole.

Eventually there will be nothing left of galaxies such as the one above, except a tiny core. (Courtesy National Optical Astronomy Observatories.)

Some of these dead stars may still have planets orbiting them. It is even possible that one of them will be our sun, with Earth still in orbit around it. If, by chance, this is the case, we might ask: What will eventually happen to Earth? We could, in fact, extend this to all planets orbiting dead stars. It is well known that any two objects orbiting one another lose energy via gravitational radiation. This means that an orbiting planet will slowly spiral into its star. But again it takes a long time—about 10^{20} years.

At this stage the universe will be a bleak place: no light, just endless darkness with dead, burnt-out remnants of galaxies occasionally breaking the void. The concept of time will, by now, have become almost meaningless. We measure time by observing various astronomical periods. This will no longer be possible.

STAGE 4: PROTON DECAY

Although much of the matter of the universe will have fallen into black holes, protons will still abound. At one time scientists believed protons were stable, and the open universe would end with them in it. But in the mid-1970s it was shown that protons might be unstable. This prediction came about as a result of a search for a theory that would unify all of nature—a unified field theory. I talked about this earlier, but to refresh your memory I'll briefly discuss it again. As you know there are four basic fields of nature. The two most familiar ones are the electromagnetic and gravitational fields. The other two are the strong nuclear force, the force that holds nuclei together, and the weak nuclear force, which is responsible for radioactive decay of the elements.

The electromagnetic and weak forces were brought together into the electroweak theory in 1967 by Steven Weinberg, who was then at MIT, and independently by Abdus Salam of Imperial College in London. Scientists then rushed to bring the strong nuclear force into the fold, thereby creating a unified electroweak–strong force, or grand unified theory. This was first accomplished in 1974 by Howard Georgi and Sheldon Glashow of Harvard. Since then several other grand unified theories have been published, and it is of particular importance that all of them predict that protons should decay.

Shortly after Georgi and Glashow put forward their theory, Steven Weinberg, Helen Quinn, and Georgi showed that the proton's lifetime should be about 10^{31} years. With such a long lifetime it might seem that it would be impossible to check. After all, the universe has only been around about 10^{10} years. But this isn't the case. If we assemble 10^{31} protons, on the average, one should decay a year. In fact, if we assemble 10^{33} we should get a decay every few days. And a group of 10^{33} protons wouldn't take up as much room as you might think (about the volume of a small house). Furthermore, since protons are the same whether they are in gold or iron, we can select relatively cheap materials such as iron and water.

As expected, several experiments have been set up (the total is now about 12). One is in a gold mine in India, one under Mt. Blanc in the Alps, and one in a salt mine under Lake Erie, to mention only a few. The experiment has to be performed underground because cosmic rays affect the detectors in the same way the proton's decay products do.

So far no one has seen a proton decay, which indicates its lifetime is longer than 10^{32} years. And if it is, it may be extremely difficult to detect. This also means that the simplest grand unified theory, the one discovered by Georgi and Glashow, is ruled out.

Once it was established that the proton might decay, several physicists realized that it would have important consequences in relation to the future of the universe. Among them were Duane Dicus of the University of Texas at Austin. Along with his student John Letaw, and colleagues Vigdor and Doris Teplitz, he began investigating these consequences.

Dicus received his B.A and M.Sc. from the University of

Washington, and his Ph.D. from UCLA. His interest in cosmology began when he was an undergraduate student. "I read some of Gamow's books," he said. "But I never really thought cosmology was an area to work in. Then I got a degree in particle physics, and eventually became interested in applying particle physics to astrophysics and cosmology."

Dicus said he began working on the problem because he felt that very little work had been done in the area. Soon after beginning his investigation, he and his colleagues realized that there were two problems to consider: how the decay of the protons inside dead stars affected them, and how the decay affected the gas floating in space. They showed that the decaying protons inside dead stars would keep them warmer than their surroundings because the decay products, namely positrons, electrons, and photons, would be absorbed by the star. Dead stars would, in fact, eventually reach an approximate equilibrium temperature: about 100 K for massive ones, and

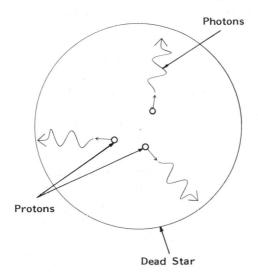

Proton decay in a dead star.

about 3 K for small ones. According to Dicus and his colleagues they would cool to this temperature in about 10^{20} years, and would remain there until all the protons had decayed in about 10^{32} years.

Dicus and his colleagues also looked at the protons of the interstellar gas between the dead stars. Its density at this stage is extremely low, and in time all of them would decay. The electron and positron density would also be low; furthermore, they would be spread throughout a universe that is perhaps 10^{20} times as large as it is today. Collisions would therefore be rare. By 10^{32} years the universe would consist of only electrons, positrons, neutrinos, photons, and a few supermassive black holes.

Other scientists also soon became interested in the effects of proton decay. In 1978 John Barrow and Frank Tipler of the University of California at Berkeley published a paper in *Nature* with the provocative title, "Eternity is Unstable" in which they showed that if the universe is at critical density the electrons and positrons would form bound pairs. In a sense these pairs would be atoms, but their dimensions would be enormous.

Don Page of Penn State University read the paper and soon realized there was a flaw in it. Born in Bethel, Alaska, Page went to William–Jewell College in Missouri, then to Cal Tech for a Ph.D. He worked with both Kip Thorne and Stephen Hawking. "I was working on my thesis when Hawking came to Cal Tech in 1974," said Page. "I was doing rate calculations and we got together and did a paper on exploding black holes and gamma rays in space. My thesis was actually under Thorne, but I interacted so much with Hawking that Thorne had me put him down as an advisor. So I actually did a joint thesis . . . it was more related to the stuff Hawking was doing than what Thorne was doing. After I graduated in 1976 I went to Cambridge University on a postdoctoral to work with Hawking. I stayed there for three years."

When Page read Barrow and Tipler's paper he quickly realized that the process they described was wrong. "They had a crude argument suggesting that electrons and positrons would

Don Page.

become bound, but they didn't say how," he said. "They seemed to have in mind a two-body interaction, but I calculated that it wouldn't lead to binding. However, when I tried a three-body interaction, I found it worked. The electron and positron are not bound initially, but a third particle comes in and carries off some of the energy and leaves them bound."

About this time Randall McKee came to Penn State as a graduate student. Needing three additional credits to graduate, he asked Page if he could do a special project under him. Page had just started doing calculations on positronium (the electron–positron "atom") so he gave some of them to McKee.

With the three-body interaction Page and McKee found that the electrons and positrons would form positronium atoms after about 10^{70} years. Once these atoms formed, the positron and electron would orbit one another, losing energy. Strangely, at this stage, they would be gigantic—as large as the present-day universe. But as they emitted photons they would spiral toward one another, annihilating in about 10^{116} years.

The argument for the existence of these strange atoms assumes the proton decays. In fact, most of the modern ideas about the latter stages of the universe assume it. And, as I mentioned earlier, there appears to be evidence that it may not decay—at least, the simplest grand unified theories have been ruled out. I mentioned this to Page and asked him what the effect of it would be in relation to his ideas. "If protons don't decay, then presumably stars will become black holes, neutron stars, and white dwarfs which will last for an enormously long time," he said. "But people have shown that if grand unified theories don't make protons decay, gravitational effects will. Instead of entire stars becoming black holes immediately you might have small collections of protons and neutrons creating small black holes within them. We know that general relativity allows large black holes. But is there some lower limit to a mass that can form a black hole?" Answering his own question he said, "You might think this mass is the Planck mass [the mass–energy at which all forces unify]. You need about 10^{19} protons for it. Or it might be possible that not that many particles are needed. Tiny virtual black holes might form, and if they do, smaller numbers of protons or neutrons would be needed."

Page admitted his ideas were speculative, but they are interesting nevertheless. Of course, we are still left with the black holes—both supergiant ones, and tiny ones. What will happen to them? This takes us to the next stage.

STAGE 5: HAWKING EVAPORATION

If protons aren't stable, perhaps black holes are. We saw earlier, though, that Hawking showed that even black holes evaporate. When quantum theory is applied to the region close to the event horizon of a black hole it is found that particles appear to be emitted. What is happening, in essence, is that the gravitational "stretching" forces (called tidal forces) are so great here that they literally pull particle–antiparticle pairs out of the vacuum. This may seem strange; after all, we usually think of space as being empty. But this is not true: it is actually a beehive of activity with particle–antiparticle pairs being created continuously. But energy is needed to create pairs. Where does it come from? Oddly enough, it is available because of a principle called the principle of uncertainty. This principle gives us a little leeway: the energy can be borrowed and if it is paid back quickly enough it won't be missed. Thus, within the cloak of uncertainty, myriad pairs are created in empty space, but almost as soon as they are created they come together again and annihilate one another.

This is, in fact, what happens near the event horizon of a black hole. But there is a difference in this case compared to empty space: there are strong tidal forces here and they separate many of the particles and antiparticles before they have a chance to annihilate one another. Thus, some of the particles fall into the black hole, but some escape, as does some of the radiation created in the annihilation process. From a distance, therefore, the black hole will look like it is emitting particles and radiation. This means, in effect, that they would not have a surface temperature of 0 K as would be expected; they would be "hot."

Hawking showed, however, that the process was negligible for stellar-mass black holes. Their temperature would only be a few millionths of a degree. For tiny black holes, though, the process would be important. They could have exceedingly high surface temperatures. Getting back to the massive black holes, we find that although they radiate slowly, they do radiate. And this means that they lose energy, and therefore become smaller. But smaller black holes radiate faster. And so it goes: the black holes get smaller and smaller and radiate faster and faster until in the last seconds of their life they explode.

This means that even massive black holes will eventually disappear as a result of Hawking radiation, but it will take an exceedingly long time. Stellar-mass black holes will live for about 10^{25} years, supermassive or galactic black holes about 10^{100} years. Most of the particles that come off will be photons, and therefore in its final stages the universe will consist of mostly photons, with a few electrons, positrons, and neutrinos. It will, indeed, be a dark, dreary, desolate place.

THE UNIVERSE AT CRITICAL DENSITY

I mentioned earlier that huge positronium atoms would only form if the universe was at critical density. Inflation, in fact, tells us that it should have this density—in other words, it should be balanced between being open and closed. If this is the case, it will, of course, expand forever—but barely.

Because the universe may be at critical density it is worth looking at this case a little closer. It is not actually much different from the case we have been discussing, but there are small differences. The expansion of the universe will gradually slow down as gravity weakens. Eventually it will almost stop—distant galaxies will barely be separating—but it will never completely stop. The processes that were discussed earlier will still take place. Protons will decay and black holes will evaporate. Huge positronium atoms will form, and they will spiral into one another emitting photons.

CHAPTER 14

Fate of the Closed Universe

In the last chapter we looked at the fate of the open universe: a cold, thin sea of radiation—essentially, a void. Let's turn now to the closed universe. If it is closed its expansion will eventually stop and it will collapse back on itself. The first question we are likely to ask is: How much longer will it expand? The answer to this depends on its density. If it is just barely over the critical density it will continue expanding for trillions of years. If, on the other hand, it is three or four times as great as the critical density it will stop expanding in about 30 billion years. Let's assume for simplicity that it is twice the critical density; in this case the turnabout will occur in about 50 billion years.

If this is, indeed, the case, astronomers will notice little difference in the universe for billions of years. The galaxies will continue to expand away from us, but they will gradually slow down. Meanwhile, stars will continue to be born and die. Anything that occurred in the early stages of the open universe will still occur. The larger stars will collapse to neutron stars and black holes, the smaller ones to white dwarfs. And since red dwarfs are so long-lived many of them will still be around in 50 billion years. Our sun will, of course, not be here, nor will any of the giant or even medium-sized stars. A few planets will have been lost to space, but most will still be in orbit around their star.

Galaxies will therefore consist of white dwarfs, neutron stars, black holes, and other smaller objects such as planets and asteroids. All the gas will now be used up so no new stars will be forming. Also, the cosmic background radiation will have cooled from 3 K down to about 1½ K.

Finally, the expansion will stop and collapse will begin. The process is similar to a ball that is thrown upward. We know it eventually stops and begins to fall back to Earth. And just as the ball falls slowly at first, gradually gaining speed, so too will the universe. For millions of years there will be little evidence of the reversal, but eventually astronomers will notice that spectral lines are blueshifted, rather than redshifted.

As the universe continues to collapse the galaxies will gain speed, falling faster and faster. Finally, clusters will begin to merge. We know that the universe, on a large scale, is made up of clusters of galaxies. Our system, the Milky Way, is part of a system of about 30 galaxies, known as the Local cluster. On a larger scale, the Local cluster is part of a supercluster called the Virgo supercluster. And it will be these superclusters that will come together first, followed by the clusters themselves. Finally, in about 100 billion years the entire universe will be a roughly uniform distribution of clusters.

From here on until the "big crunch" the action will be faster. Individual galaxies will begin colliding. If anyone is on Earth (a very unlikely possibility) they will see the Andromeda galaxy growing larger and larger in the sky. Finally, it will merge with our system, as will other galaxies in our group. Oddly enough, though, these collisions will not be dramatic events: individual stars will not collide. They are so far apart that they will simply glide past one another. If there were any gas left in the galaxy, it would collide, but at this stage it is unlikely there will be.

From Earth the sky will now be magnificent: it will be dotted with exceedingly bright stars. It will still be black at this stage, but will not remain so for long because of the cosmic background radiation. This radiation has a temperature of 3 K today, but as the universe collapses it is compressed and heated. Furthermore, starlight is compressed and it also heats. At first the sky will just take on a ruddy glow. But as the temperature of the

A cluster of galaxies. Eventually all such clusters will come together. (Courtesy National Optical Astronomy Observatories.)

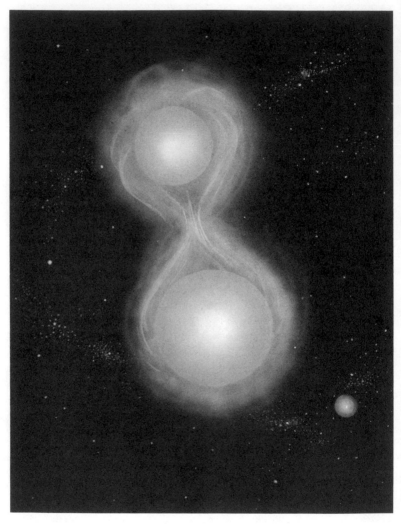

Interacting white dwarfs. Their outer layers are evaporating off to space.

The merging of two black holes in the distant future.

radiation increases it will get brighter. Within days it will be up to a few thousand degrees, then within hours it will soar to 6000 K, the approximate surface temperature of the sun. The entire sky will suddenly be ablaze—as bright as the surface of the sun.

At this stage it is unlikely there will be any stars left— perhaps a few red dwarfs. But there will be white dwarfs and neutron stars in abundance, many of them still hot. But the temperature of space is now hotter than their surface, and they cannot release their radiation. Their outer layers will therefore begin to boil off to space, and they will begin to look like comets with evaporation tails trailing behind them. When the temperature finally gets to a few million degrees whatever remains of them will explode and dissipate off to space. Then there will be nothing left but black holes and particles.

The universe will now be a hot plasma—a soup of charged particles. Mixed with this soup will be energetic radiation in the form of X rays and gamma rays. The only large objects left will be black holes, and since the universe is now much smaller they will be much closer together. Furthermore, they will now be growing rapidly, fed by the thick soup around them. And as they come closer together, they will begin cannibalizing one another. Smaller ones will be absorbed by larger ones, so their numbers will go down, but the ones remaining will become gigantic.

APPROACHING THE SINGULARITY

The universe is now opaque, like a thick, hot fog. Aside from the black holes there are only particles: neutrons, protons, electrons, positrons, neutrinos, and photons. But soon even the protons and neutrons will disappear. They are made up of "bags" of quarks—three to a bag. And when the temperature reaches approximately a million-million degrees (10^{12} K) the bags will rip open and the quarks will begin to spill out. When this happens there are no protons or neutrons left—only quarks. And as the temperature continues to soar the first of the four fundamental forces of nature will merge. The weak and the electromagnetic fields will come together creating a single, unified electroweak field.

Heavier and heavier particles are now being created. The universe is like a gigantic accelerator. On Earth we build such accelerators to produce new massive particles, but here nature is producing them throughout the universe. Soon the temperature is up to 10^{20} K, then 10^{25} K. The third of the basic forces of nature then merges with the electroweak field and we have a grand unified field. Only gravity remains outside the unification.

The processes that are taking place now are the same as those that occurred when the universe was being created, but in reversed order. There is, however, a major difference. The universe now contains large numbers of massive black holes, many in the process of colliding. In the early universe tiny primordial black holes were likely created within the first fraction of a second, but they were not as massive as these.

A critical point comes at 10^{-43} second before the "big crunch." All forces of nature have now come together: gravity has merged with the grand unified field. Furthermore, there are fewer types of particles. In fact, there may only be one type of particle. With one force and one particle the universe is much simpler than it was earlier. But for scientists there is now a major problem: because the universe is so small, quantum forces are as large as gravitational forces. Once this happens Einstein's theory of general relativity, the theory we are using to follow the big bang, breaks down. It is not a quantum theory, and therefore no longer gives correct answers. So in the last tiny fraction of a second we are not certain what happens. If the universe shrinks to a point, it will have infinite density; it will be a singularity. But is a true singularity actually reached?

Scientists are still not sure because of the breakdown of the theory. One possibility is that the universe "bounces."

BOUNCE

If the universe does bounce it will stop just before it becomes a singularity, and reexpand. It will then go through the same stages it just passed in its collapse. And since a closed universe remains closed for all time, if it bounced once it will bounce again. This means it will bounce forever. We refer to such a model as an oscillating universe. (The mathematics of this model was first examined in the 1930s by Richard Tolman of Cal Tech.)

One of the reasons scientists became interested in this model was that it gets around what some consider a serious problem: a beginning to the universe. If the universe cycles, it will cycle forever, and therefore there was no beginning—or so it seems. Recently, however, theorists have shown that things aren't this simple. Starlight is accumulated from cycle to cycle and, because it is a form of energy, it causes each cycle to be longer than the one before it. Looking back in time, then, we see shorter and shorter cycles, and eventually we come to a time when there were no cycles. According to calculations there can be no more than 100 cycles before our present one. And this, of course, means our problem is not solved.

Tommy Gold of Cornell University has gotten around this in an ingenious but highly speculative way. He assumes that there is a time reversal when the universe gets to maximum expansion. The universe will then get smaller and everything will occur in a reversed order. We will get younger instead of older. But we would certainly notice this. So, how does this help? According to Gold, our mental processes would also be reversed so everything would appear normal to us—we would age in the usual way. Gold's universe would oscillate back and forth between the big bang and maximum expansion, and cycles would not have to be longer than those that came before them.

QUANTUM COSMOLOGY

But will the universe bounce? For an answer to this we have to know what goes on once the universe reaches the quantum realm (region where quantum forces are as large as gravitational forces). The theory we use is a cosmology based on Einstein's theory of general relativity. It explains the events that come after the big bang, and gives us reasonably satisfactory answers. Slight modifications have been introduced lately to overcome some of the problems of the theory, but in general, it's an excellent theory. It does not, however, explain everything under all circumstances. This shouldn't surprise you, though; the same thing happens in Newton's theory. We know, for example, that it explains all macroscopic phenomena: colliding balls, motions at moderate speeds, and so on. But once you get into the realm of atoms and elementary particles it breaks down. It no longer gives the correct answer and must be replaced with quantum theory.

This also happens in the case of general relativity. When the universe enters the quantum realm we find that it also breaks down. To get a correct answer we need a quantized theory of general relativity, or more explicitly, a quantized cosmology. And so far we don't have one.

At the present time modern theoretical physics consists of two fundamental theories: quantum theory (or more exactly, quantum field theory) and general relativity. Each has developed separately; one explains the very large, one the very small. And there appears to be no overlap between them; they are separate theories. But why should nature need two different theories to explain it? It seems that one would be sufficient.

Most physicists agree that this is one of the most important problems of science today. And, needless to say, there have been countless attempts to formulate such a theory. All the "greats" of physics—Einstein, Dirac, Heisenberg, Pauli, Eddington—have tried their hand at it, but it eluded them all. And it still eludes us. Indeed, if the truth be known, surprisingly little progress has been made. Despite the years of work we still are not sure what form such a theory would take, or even what gravitational quantization means, or involves.

The major problem is that the gravitational field is different from the other three fields of nature, and because of this an entirely different approach may be needed. So far there have been two major approaches. The first, the one used by Einstein, is to attempt to generalize general relativity; in other words, to try to broaden it to include the other fields of nature, and at the same time hope that the broadening process somehow introduces quantization (and explains the elementary particles). In recent years most of the effort has gone into the other approach. We have a tentative unification of three of the fields— the grand unified theory. What is needed is to expand it to include gravitation.

One technique has been to increase the number of dimensions of the theory. We know that we live in a four-dimensional world: three dimensions of space and one of time. And, of course, most theories—general relativity included—are four-dimensional. It is quite easy, however, to increase the number of dimensions in a theory; it only involves changing the equations slightly. Of course, in the end, after the calculations are done, we have to come back to our world of four dimensions. The first person to do this was Theodor Kaluza, who generalized Einstein's general theory of relativity to five dimensions in 1921. He found, to his delight, that the resulting theory included the electromagnetic field. It was later extended by Oskar Klein, who explained why the fifth dimension was not observed. He showed it was curled in the form of a cylinder that was so small it would not be seen. The theory is now referred to as the Kaluza–Klein theory. Higher dimensional theories today include 10 or 11 dimensions, but the principle is the same.

Another approach to the problem involves a theory I talked about in an earlier chapter, namely supergravity. Scientists have shown that supergravity both simplifies the particles of nature, and incorporates general relativity. Furthermore, it is a quantum theory. For a while it seemed as if it might be the theory that we were looking for. But when it was examined in detail it too was found to be lacking.

Then came an important breakthrough. Scientists had, for years, been working on a theory based on two-dimensional "strings"—but it never seemed to work properly. Then in 1984 it was shown that a variation of the theory could do things that supergravity couldn't. In this theory, called superstring theory, scientists deal with tiny strings—only a billion-billion-trillionth of a centimeter long. They are different from the usual strings of our experience in that they have a tremendous amount of energy associated with them. They can be either closed, like an elastic band, or open. And they can interact with one another. When two touch, for example, they can come together, creating a single string.

What is particularly important, however, is that they vibrate. Their vibrational modes can represent all the known particles, and can explain the four forces of nature. Furthermore, the theory contains general relativity. At the present time it seems to be the most promising approach to bringing together general relativity and quantum field theory. And once we are able to do this we may know whether or not our universe will bounce.

THE MANY-UNIVERSE CONCEPT

While we're in a speculative mood let's look at some of the other speculative ideas that cosmologists have. First of all, we may not only have many universes existing one after the other (i.e., cyclic), we may have several existing at the same time. The idea of parallel universes has been around for several years. It first came up in quantum theory when Hugh Everett III suggested it in his 1957 Princeton University doctoral thesis. Few took notice of the idea at the time, though—it seemed too far out. Everett explains some of the problems of quantum theory by assuming the universe splits into innumerable replicas with each new universe again splitting. This got the ball rolling. And although Everett was not thinking about cosmological models it got several cosmologists thinking about the possibility.

But how does this fit in with our current ideas of creation? Let's begin with creation itself. One idea is that our universe began as a "quantum fluctuation"—in other words, it was created spontaneously, in the same way particle pairs are created spontaneously out of the vacuum. If this is so, it seems that if one universe could be created, many could be created. Andrei Linde of the USSR put forward a theory of this type in 1983; he referred to it as chaotic inflation. In his theory he presents arguments for many self-reproducing universes, some of them larger than ours, some smaller. He visualizes a "space-time foam" of incredible energy. Independent, separate universes, according to Linde, emerge from this foam, each expanding at its own distinct rate. And each having its own distinct number of dimensions.

A variation on this idea is that universes are created within other universes. This might mean that our universe began as a quantum fluctuation in some other universe. And other universes are ready to pop into existence out of ours—or perhaps even have in the past.

All of this sounds rather crazy, but scientists do talk seriously about such things. They have even considered the possibility of an advanced civilization creating a new universe in the laboratory. It goes without saying that this wouldn't be possible for millions of years—if at all. Anyway, so much for many-universes, and I hasten to add that if you found this section speculative, you may be mind-boggled by the next.

THE ANTHROPIC PRINCIPLE

John Wheeler of the University of Texas and, independently, Stephen Hawking of Cambridge University have taken a close look at the region where general relativity breaks down, in other words, the quantum region, and they have come up with some interesting results. They believe that once the universe enters this region a "fuzziness" develops. Their belief is based on a principle of quantum theory called the Uncertainty Principle. This principle tells us we cannot simultaneously measure such things as the position of a particle and its momentum. If we narrow in on one, the other becomes fuzzy.

Using this they showed that if the universe collapsed beyond a certain point (a density of approximately 10^{40} g/cc) the fundamental constants of the universe, things such as mass and charge of the electron and the gravitational constant, may be lost. This means that if the universe emerges from this region, it will emerge with different fundamental constants. If so, this will likely affect its rate of expansion, so it will end up being a different size. In fact, the argument I mentioned previously about each cycle having to be longer than its predecessor no longer applies here; the cycles can be any length. And this presents a problem in relation to life. Some of the cycles may be quite short—only a few billion years long. But we know life took at least 3 billion years to develop on Earth, and it was another 2 billion before higher intelligence and civilization appeared. Therefore, any cycles shorter than this would likely be void of life.

Of particular importance in relation to this, however, is a discovery made a few years ago by Brandon Carter of Cambridge University. He was examining the effects of changes in the fundamental constants of the universe on the stars (their makeup and evolution) and found that if these constants were changed, even slightly, many types of stars would not exist. For example, if they were adjusted in one direction all stars would be small—mostly red dwarfs. But we know that the elements (with the exception of a few of the light ones) are produced in massive stars. Therefore planets such as ours could not exist. On the other hand, if they were adjusted the other way, only giant stars would exist—none as small as our sun. And we know that stars like our sun are the best candidates for life-supporting systems. This means that life would be impossible in a universe that had fundamental constants that were very much different from the ones our universe has. This seems to imply that life is tuned to the fundamental constants that it now has. If they were much different there would be no life—and we would not be here. This idea sums up what is known as the Anthropic Principle.

But what effect do evolution and the various "coincidences" that occur in our system have on the idea that we are "tuned" to our present fundamental constants? Let's begin with the coincidences. The tilt of the axis of Earth is ideal; if it were tilted differently (smaller or larger) weather conditions on Earth would be far less favorable. Also, without the ozone layer, it is unlikely life would have formed; UV radiation from the sun would have destroyed any life that started to form. And without a magnetic field we would be struck at a far greater rate by cosmic rays. Furthermore, our gravitational field is ideal for us. And certainly our distance from the sun (93,000,000 miles) is optimum. Move Earth outward or inward slightly and our weather would change dramatically. On the basis of this, it seems as if Earth was made for us. And if things were different—regardless of the fundamental constants of nature— we wouldn't be here.

Looking at things a little closer, though, we see that this isn't necessarily so. I've skipped a key element: evolution. If things would have been different, the first organisms would have evolved differently: they would have adapted to the new environment. If gravity was strong, life forms would have developed larger muscles; if we were closer to the sun, they would have developed some mechanism for cooling themselves. And so on.

But of course we know there are limits. We certainly couldn't have evolved if conditions were drastically different. In the same way there is a certain amount of variation in the fundamental constants that we could stand—but not much. For the most part, the universe does, indeed, seem to be tailored to our needs.

CHAPTER 15

The Future of Civilization and Life

FATE OF EARTH

We have looked at the fate of both the open and closed universes. But a more pressing problem for us is: What is in store for mankind? What is the fate of Earth? This, strangely enough, is not an easy question to answer. We can predict the future of the universe much better than we can predict the future of civilization and life. There are so many factors, so many different things that could happen when it comes to civilization and life that it is difficult to make accurate predictions. Nevertheless, there are only a finite number of possibilities (hopefully), so let's look at some of them.

One of the most serious short-term threats is overpopulation. Writers frequently ignore this when they write about the future of Earth. But it can't be ignored. As strange as it may seem, our planet is almost full now, in the sense that it is not able to support many more people. Yet the population explosion goes on. We don't hear much about it in the news anymore, mostly because it has leveled off slightly. Nevertheless, it's still with us: Earth's population is doubling every 35 years. In fact, in some countries it is doubling considerably faster than this (it's 20 years for Mexico). And even if we somehow manage to extend this to, say, 50 or 100 years, we're still not out of the woods. We will still, in theory, expand to an infinite population late in the 21st century.

And this isn't all. The poorer nations of the world are now developing and are therefore using more energy, and we ourselves are increasing our rate of energy consumption. We will therefore, in theory, need an infinite amount of energy sometime in the next century. We know, of course, that something will happen to stop this explosion. But any way you look at it there is a problem.

Aside from the population problem there's also the threat of nuclear war. War has raged literally since the beginning of civilization. At one time it was only tribe against tribe, but as technology developed it became nation against nation, then many nations against many nations. Most wars of the past were started over petty differences, and while people are likely to be a little more careful now that we have weapons that can kill millions in seconds, it is likely that the fighting will go on.

But even if we survive for a long time without nuclear war, other possible calamities await us. One is related to the greenhouse effect. In this effect, radiation from the sun becomes trapped between the surface of a planet and a cloud layer above it. This is caused because its wavelength is changed when it strikes the surface and it is no longer able to penetrate the clouds. Of course, some still gets out; if it didn't the temperature would soar to infinity. This effect is quite serious in the case of Venus; it increases its temperature by several hundred degrees. And, of course, this creates conditions that are hostile to life—to say the least.

But the greenhouse effect doesn't always create disastrous conditions. In the case of the earth it is quite beneficial. It increases Earth's average temperature by about 60 degrees, which makes it a hospitable place for us. But as we pollute our atmosphere, less radiation will leave and temperatures will increase. Eventually it may become inhospitable.

Oddly enough, there is another effect that is countering this slow heating. Before you give a sigh of relief, though, I should mention that it will only work for the next few hundred years. We know that our sun undergoes an 11-year sunspot cycle, with large numbers of spots occurring at certain times, and few, if any, $5\frac{1}{2}$ years later. On a longer time scale, about 80 years, it undergoes another cycle, with few sunspots even at sunspot maximum. And to some extent the weather of Earth is affected by this cycle.

More important, though, is that over even longer cycles there are periods of about 100 years where there are almost no sunspots. This phenomenon was first discovered by Gustav Spörer in the late 1800s. He noticed that there were very few sunspots between about 1600 and 1700. British astronomer Walter Maunder verified the discovery and the era is now known as the Maunder minimum. A similar earlier era is named for Spörer. Between these eras is what is called a grand maximum—a cyclic region of many sunspots. We are, in fact, now on the tail end of the grand maximum that followed the Maunder minimum.

But what is important about this? Again, it's weather. Looking back through weather records we see a significant cooling during the Maunder minimum. There was what is sometimes called a "mini ice age" throughout Europe—a period when the average temperature was about 5 degrees below normal. Furthermore, there were severe drought conditions throughout parts of the world.

Five degrees may not seem like much, but over many years it would have a serious effect on crops. And, of course, these mini ice ages are only a preview of what may come in even longer cycles. We know that there have been significant ice ages in the past. Much of North America was covered with ice during these eras. Does this mean we're in for another ice age? There's certainly no reason to believe we've seen the last of them.

But there are other things that will create problems long before another ice age hits us. So it's hardly worth worrying about. Getting back to the problem of energy resources, let's consider how it could be overcome. Our major source of energy at the present time is the sun. But we are picking up only a small fraction of its total energy; most of it is lost in other directions in space. And since it is such a valuable resource it seems likely that in the not-too-distant future we will build large mirrors to harness some of the lost energy. Indeed, in the distant future we may even try to harness all of it. Freeman Dyson suggested a way of doing this a few years ago. He envisioned a large sphere around the sun that collects all its radiation. Admittedly, this sounds like science fiction, but at our present rate of consumption we'll need this amount of energy in about 500 years.

So much for pessimism. Let's be optimistic for a while and assume that life survives for millions of years into the future. We will likely have traveled to the stars by then, but we'll leave the details of that for later. For now let's look briefly at what will eventually happen to Earth. The major catastrophe will come in about 3 billion years when the sun begins to expand. As its surface moves outward, our oceans will boil off to space, and our atmosphere will evaporate. As its expansion continues the sun will encompass Mercury, then Venus. From Earth at this point it will look like a gigantic red sphere, taking up much of the sky. But just before it swallows Earth its expansion will stop. Then slowly it will begin to shrink, and temperatures on Earth will moderate. Finally, it will become a white dwarf, and from Earth it will only be a bright star in the sky.

THE EXODUS TO NEAR SPACE

Let's back up now and look in more detail at what will happen in the near future. As our planet becomes overpopulated and energy resources dwindle we'll no doubt expand into the solar system. I should stress, though, that this will not be a solution to our population problem. Even if we had spaceships capable of holding hundreds, that took off regularly—say every few days—it wouldn't be a solution. Earth's population is expanding at a much faster rate than this.

I asked Freeman Dyson how long he thought it would be before we were forced into space because of overpopulation and energy problems. "Overpopulation and energy resources will be problems," he said. "But it's not a question of being forced. It's a question of wanting to do it, and we will want to. I'd say it will begin in about 100 years. But that's only a guess."

The first expeditions will no doubt be on a small scale: the

A lunar base in the 21st century. (Courtesy NASA.)

setting up of space stations for scientific purposes. (We have, of course, done this to some degree already.) Later, large-scale stations will be constructed that can support hundreds or even thousands of people. Gerald O'Neill of Princeton University has spent several years thinking about this problem. He feels that ring-shaped cylinders that spin to simulate gravity are our best bet. According to O'Neill, the sides of these cylinders should be made of alternate strips of land (soil) and glass (or other transparent material) so that shutters can be used to simulate day and night. He feels that up to 200,000 people could live in such a colony.

And, of course, there are also the other planets (and moons) of the solar system. The major problem here is lack of an atmosphere, or at least a proper atmosphere. This means that col-

Constructing a space station in the 21st century. (Courtesy NASA.)

onies will have to be housed in large airtight domes. In time, though, we might be able to do away with these domes through what is called terraforming. This is the altering of atmospheres, climates, and so on, so that they approximate those that exist on Earth. Certainly it would take a long time to do something like this, but it is possible.

Consider the atmosphere of Mars. It is quite different from ours: no free oxygen, no water vapor, and its pressure is only about 1 percent of ours. Furthermore, deadly UV light from the sun reaches the Martian surface. But all this could be changed. Temperatures, on the average, are well below those of Earth, so we would have to warm it. And we would have to get water flowing on its surface. The water problem may not be serious—since the pole caps are composed mostly of water ice, and there

Space station near Saturn. (Courtesy NASA.)

appears to be a layer of permafrost under the surface. With its thin atmosphere, however, even if we could melt some of the ice, water would not flow on the surface. Our first job, then, would be to increase the atmospheric pressure.

How would we do this? One way would be to place a large mirror in orbit around the planet; sunlight could then be directed at the pole caps. Although these caps consist mostly of water ice, there is a thin layer of frozen carbon dioxide (dry ice) overlaying it. With enough heat this layer could be sublimated into the atmosphere, thereby increasing its pressure. And therefore when the ice began to melt some of it would appear as water and flow on the surface. Most of it, however, would go into the atmosphere as water vapor, but this would build up its pressure even more. And in time as the pressure increased, and clouds developed, more water would flow. Finally a greenhouse effect would begin which would help heat the surface.

Of course, for life we also need oxygen. And in this case the solution is again relatively simple. It was found during the Viking experiments that when water was added to the Martian soil, oxygen was given off. This means that as water flows over the land, oxygen will be released to the atmosphere. Furthermore, we could speed things up a little by adding specially tailored algae that convert carbon dioxide to oxygen. Strains of algae have already been produced that thrive in a simulated Martian atmosphere.

Once oxygen was introduced, an ozone layer would begin to build, which would protect the surface from the deadly UV radiation. Thus, we would have water, an appropriate atmosphere, and moderate temperatures—everything we need. And indeed it sounds as if it would be relatively easy to do. But in practice it would require considerable engineering ingenuity, and would probably take at least 500 years. Nevertheless, it could be done.

Venus could also be terraformed. There are more difficulties in this case. Again we would have to bring water to the surface. Since there is very little water on Venus, it would have to be transported from, say, one of the outer planets. We would also have to cool the atmosphere and get rid of most of the carbon dioxide. And perhaps the most difficult feat: we would have to change the rotational period of the planet. At present its rotational period is much too long (243 days); we could never get conditions approximating those on Earth with such a long period. Estimates of the time it would take to do this range from about 500 up to 1500 years or more.

The outer planets, because they are gaseous, would not be useful as abodes for life. But they would be extremely useful as sources of materials for terraforming other planets and moons.

FLIGHT TO THE STARS

Once the resources of the solar system are used up we will no doubt look to the stars. Trips to the nearest stars will be attempted first. This will, of course, not happen for a long time because of the distances involved. The distances, themselves, wouldn't be such a problem if it weren't for the fact that we are limited in how fast we can travel. According to relativity we can't travel faster than the speed of light. This means that if a star is 20 light-years away there is no way we can get to it in under 20 years (at least none we know of at the present time). Because of the accelerations and decelerations needed, it would, in fact, take us considerably longer than 20 years, even if we had the technology to travel at a speed near that of light.

Although it is difficult to imagine how far away 20 light-years is, we can get a feeling for it by considering a beam of light. A ray from our sun (distance: 93 million miles) takes 8 minutes to get to us. If it continued on to Pluto, 4 billion miles away, it would take $5\frac{1}{2}$ hours to get to it. And if it continued to Alpha Centauri (our nearest star), 7000 times farther out, it would take 4.3 years. Finally, to get to our star at 20 light-years it would take 20 years. And yet, astronomically speaking, 20-light years is a small distance; there are galaxies that are millions and even billions of light-years away.

So, although we can travel to the stars in science fiction, we are far from being able to do it in practice. Pioneer 10, which was launched in 1972, recently left our solar system at a speed of about 35,000 miles per hour. Voyager is also on the way out at a slightly higher speed, but still an insignificant fraction of the speed of light. At this rate it would take these spacecrafts over 80,000 years to reach Alpha Centauri (if they were headed in that direction—but they're not).

Eighty thousand years may seem like a long time (and indeed it is) but compared to the time it would take us to get to more serious candidates for life, it is small. One of our best nearby candidates, Tau Ceti, in the constellation Cetus, is 12.2 light-years away. And if present trends are any indication the first really interesting stars may be hundreds of light years away.

So that we can appreciate some of the difficulties let's consider a flight to Alpha Centauri. It is 4.3 light-years away. Upon taking off from Earth we would have to accelerate to a speed close to that of light. That would take at least a year, but then we could coast. As we approached Alpha Centauri, however, we would have to decelerate for another year to get to a low enough speed so we could observe it, or perhaps land on an associated planet (if one existed). Our trip back would be similar: we would have to accelerate to a speed close to that of light, coast, then decelerate as we approached Earth. Calculations show that the least time we could complete the trip would be about 11 years.

But in practice it isn't time that is the major problem, it's the amount of fuel that would be needed. A convenient measure is what is called the "mass ratio." It is the ratio of the mass of the starting rocket to that of the payload (i.e., the mass when the star is reached, or when they get back to Earth again). You may remember seeing the effect of this in relation to the Apollo mission to the moon. The mass ratio in this case is the weight of the rocket as it takes off compared to the weight of the tiny capsule that is fished out of the ocean.

The mass ratio needed to get to Alpha Centauri at a speed of 99 percent that of light and return is approximately 1 billion. In other words, for every ton we want to get back we need a billion tons of starting fuel (since most of the weight is fuel). At that rate the rocket would have to be larger than the Empire State Building.

There is a way around part of this problem. Much of the fuel is used in escaping Earth's gravitational pull. Therefore, if we construct the spaceship in space using materials from, say, asteroids or moons we would not need nearly as much fuel.

Actually, there's something else that may eventually help us, but it depends on the availability of fusion reactors—and, so far, we haven't been able to overcome the problems that would allow us to build one. Let me begin by briefly discussing the fusion process itself. Fusion is a joining or fusing of nuclei (in the sun four hydrogen nuclei fuse to give a helium nucleus). It is achieved by supplying enough energy so that the nuclei get close enough to fuse. In the hydrogen bomb (which also depends on fusion) the hydrogen atoms are brought close enough by encasing them around an atomic bomb, then detonating the atomic bomb.

The main advantage of fusion is that it requires light elements such as hydrogen, which are abundant and cheap. But there is a containment problem. The temperature at which fusion takes place is so high there is no known substance that can contain the fusion products. In other words, when the reaction occurs, all known materials melt and the fusion products spill out. There has been a tremendous effort in the last few years to build a "magnetic bottle" that can contain the reaction. And there have been some successes, but we've still got a long way to go.

But even if the containment problem is overcome we still have to ask ourselves how we will carry such a large amount of fuel. In fact, there's a kind of catch-22. The bigger the rocket the more fuel that is needed, but it's the fuel that makes up most of the weight of the rocket. There is, however, a way of getting around carrying the fuel. Several years ago American physicist Robert Bussard suggested that hydrogen could be picked up along the way. Space contains ample amounts of it.

Bussard's technique may overcome the fuel problem, but even if it does there are still others. To get to exceedingly high speeds we have to accelerate, and the human body can take only so much acceleration—about 15 g's (g is the acceleration of gravity of Earth), and that only for a few seconds. This means that we may be restricted to coasting speeds of about 10 percent that of the speed of light. And therefore it would take several generations to get to most stars. People would have to spend their entire lives in spaceships, and I'm sure few would volunteer. One way around this would be to put them in some sort of deep sleep, or perhaps freeze them. But since we know little about the effects of this I won't say anything about it.

There's another problem that would have to be considered on long flights. On any flight there is a coasting phase during which the astronauts are weightless. On short flights, say, within the solar system, this phase would not last long and therefore would cause few problems. But in a flight to a star, which would involve most if not all of a lifetime, it would be a serious problem. If we were suspended in zero gravity for a long period of time our muscles would weaken, and when we were again in a gravitational field we would not be able to stand up. Because of this we would need an artificial gravitational field in the spaceship. This could be created by accelerating the rocket so that it duplicated the acceleration of gravity on Earth. But it would be difficult to do this on all flights.

And finally, if spaceships are to travel to the stars, they will have to be large—large enough to support hundreds of people over hundreds of years. The support system of such spaceships staggers the imagination. The air would have to be recycled and precautions would have to be taken so that none escaped. If only a small amount escaped each year, it would be disastrous. Water would also have to be cleaned and recycled. But perhaps the most difficult problem of all would be the people themselves and the psychological problems that space travel would inevitably create.

When starships finally encounter planets that appear habitable the occupants will not be able to just step out onto the surface. It's unlikely that the atmosphere will be appropriate. Large airtight domes will have to be used until the planet is terraformed. And this, of course, could take hundreds of years.

Despite the difficulties we may one day get to the stars. But what will we find when we get there? A question that immediately comes to mind is: Will we find other civilizations? Let's consider this.

OTHER CIVILIZATIONS

If you look at the night sky through a telescope you see millions of stars. With such a large number, the chances that there is at least one advanced civilization seems to be overwhelmingly high. And for many years most astronomers were convinced that there were many extraterrestrial civilizations—some of them likely much more advanced than us. But recently a number of scientists have come to the conclusion that this may not be the case. They are convinced that we are the only advanced civilization in our galaxy, and they give some pretty convincing arguments for their belief.

Let's begin with the point of view that there are a large number of civilizations. The main reason for this belief, as I just mentioned, is the large number of stars in our galaxy—200 billion. Furthermore, it also seems reasonable that the process that gave rise to life on Earth also gave rise to life elsewhere.

A formula giving a statistical estimate of the number of civilizations in our galaxy was developed in 1965 by Frank Drake of Cornell University. I will not go into the details of the formula, mentioning only that it involves estimating such things as the number of planets per star, the number of planets likely to be in the life zone of the star, the fraction of planets in the life zone likely to develop life, and the lifetime of a typical civilization.

There is considerable uncertainty in arriving at estimates for each of the above, so the numbers we get from the formula depend on how optimistic we want to be. They vary all the way from a few civilizations up to about a million. This is useful, however, in that for a given number we can calculate how far away a civilization is likely to be. If there are, say, a million worlds like ours out there, calculations show that the nearest one should be about 100 light-years away. Even for this optimistic case, though, it would be difficult to communicate with them—it would take 200 years for a return message. But if there is, indeed, a civilization that close we might be able to detect it right now. How? Let's assume they have followed our route to higher technology. If so, they would have developed TV stations (and high-frequency radio) and the stray radiation from these stations would be expanding out around them, just as it is expanding out around Earth now. If it has expanded for 100 years we could detect it.

On the other hand, if there were only 100,000 like us in our galaxy the nearest one would be at least 1000 light-years away. In this case it would take 2000 years to get a message back— hardly worth the effort. If you're an optimist, then, you're likely convinced there are a large number of civilizations in our galaxy. But there are scientists who are quite skeptical of this. Frank Tipler of Tulane University is one of them; he is convinced that we are the only advanced civilization in our galaxy. He bases his conclusion on two things: the small probability that advanced forms of life will evolve, and the high probability that if they did the entire galaxy would be colonized.

He points out that we are, even now, capable of sending probes to the stars. In fact, we have already sent two: Pioneer and Voyager. Granted, it will take at least 80,000 years before they encounter any stars, but when considered in the proper perspective this is not really that long. On the other hand, in the not too distant future we should be capable of sending probes at much greater speeds, say 10 percent the speed of light. In this case it would take only 45 years to reach Alpha Centauri.

But as I said earlier, we have to ask ourselves how many would volunteer for such a trip. Not many, I'm sure. Tipler realizes this and has suggested an alternative that he believes future generations will prefer. Our computer technology is advancing rapidly and it is likely we will have computers in the next few decades that will be capable of reproducing themselves. In fact, such a computer could be designed so that it would not only be capable of reproducing itself but could also build an entire spaceship.

Tipler believes that advanced civilizations would send probes of this type to the planets of nearby stars, and have them use the raw materials to reproduce themselves. Better still, they might produce several copies; these copies would then be projected at other stars. Each of them, in turn, would produce several copies, which would again be sent out. It's like the old chain letter scheme.

The problem with such a scheme is that it usually isn't long before the chain is broken. Tipler argues, however, that even with a small percentage of the probes completing their mission the entire galaxy would soon be colonized. In fact, a simple calculation shows that it would not take much longer than a few million years. Yet our galaxy is billions of years old. If this is, indeed, the case, we have to ask, as Enrico Fermi did many years ago: "Where are they?"

Tipler's arguments, not unexpectedly, have been severely criticized. Frank Drake points out that the probes would have to be extremely massive in order to carry with them all the materials needed to initiate a successful mining operation and refining factory. And, of course, the more mass, the greater the required energy to accelerate it to, say, 10 percent the speed of light.

Michael Papagiannis of the University of Boston is also convinced there should be a wave of colonization throughout our galaxy, but he has an alternative suggestion as to how the colonizers would travel. He believes that advanced civilizations will eventually take to large artificial orbiting space satellites. In time some of these satellites will move away from their star, and, according to Papagiannis, even if they traveled at only 1 to 5 percent the speed of light, they would colonize the entire galaxy in a relatively short period of time. He allows 500 years for a trip of 10 light-years and 500 years for the establishment of a colony and shows that it would only take 10 million years to populate the galaxy.

Papagiannis also points out something else that is particularly important: If there are, say, one million advanced civilizations in our galaxy right now, there would have been at least a billion such civilizations in it in the past. And if there were such a large number, surely they would have populated the entire galaxy. Yet we see no evidence of them.

Earlier I talked about Dyson spheres—shells that advanced civilizations would likely use to trap all the radiation around their star. I asked Dyson if he thought it would be possible to detect them. "We should look for them," he said. "I proposed some time ago that we should look for evidence of a technology. And what do we have the best chance of seeing? The answer is simple: something that is as big as possible. So I proposed looking for infrared sources [Dyson spheres]—the typical by-product of a technology."

But we haven't found any evidence of infrared sources; in fact, some people have even suggested that this lack of evidence is telling us that there are few, if any, civilizations out there. "I think this is silly," Dyson replied. "First of all, we haven't really started looking. Secondly, I think it is silly to try to predict what other civilizations might try to do. I'm not making a prediction. I'm only saying that if you're searching for something, you're most likely to see the brightest sources. The only infrared survey completed so far is IRAS [1986]. So I think the question is still completely open."

About all we can say at this stage, then, is that we're not sure how many extraterrestrial civilizations there are. But, since we are likely to attempt to get to the stars ourselves, it might be us that eventually populates our galaxy.

One last point: It's important to note that Tipler's and Papagiannis's arguments apply only to our galaxy. Even if we are the only life in our galaxy, there are at least as many galaxies beyond ours as there are stars in our galaxy. So if there is only one civilization per galaxy, there will still be a lot of civilizations.

HOW LONG WILL CIVILIZATION AND LIFE SURVIVE INTO THE FUTURE

We have looked at the future of life and civilization in the near future, and I have made a few comments about the distant future. Let's turn now to a more detailed discussion of the distant future. As we saw earlier our sun will eventually burn out. In about 3 billion years it will start to expand. But even when it is a red giant, energy is still being generated and there is still a

A scene on Earth today.

region around it that is satisfactory for life. But it is unlikely there will be enough time for life to form. In a short time it will collapse to a white dwarf.

By the time this has happened, though, civilization will be mobile and space arks will have explored much of our galaxy. In fact, all of the "usable" stars will eventually be used up, for even galaxies run down. They have only so much gas for the production of new stars, and when it is used up star production halts. This has already happened in elliptical galaxies.

The major source of energy at this time will likely be black holes. Through what is called the Penrose process, energy can be extracted from black holes. If we project pellets into the ergosphere of a spinning black hole, and arrange for part of each pellet to fall through the event horizon, and the other part to escape, we find that the emerging part carries off some of the energy of the black hole. This energy can, in theory, be con-

The same scene millions of years in the future.

verted to electrical energy or other useful forms. Civilizations will therefore likely take up residence around black holes, living off their energy.

As the universe continues to run down, the outer regions of galaxies will evaporate off to space and the black hole at the center will become larger. Civilizations will then begin crowding closer and closer to the core. But all galaxies are not at the same stage of development; some, even at this point in time, may still be developing stars. And civilizations will no doubt leave their galaxy in search of them. This will become increasingly difficult, of course, since galaxies are continually moving apart as a result of the expansion of the universe. By then, however, other modes of travel may have been discovered. In theory, black hole travel—travel through the spacetime wormhole associated with a black hole to distant parts of our galaxy, or even to other galaxies—is possible. At the present time, though, we know little about how this might be achieved, or even if it is actually possible.

Long-time survival depends, of course, on whether the universe is open or closed. If it is closed it seems certain life can survive only up to the big crunch. In the open universe we don't have to worry about being crushed, but we have equally serious worries. If the protons decay, all life (which is made up of protons) will disappear. Dyson has suggested that life forms might eventually change structure, slowing down their metabolism rate so that they need little energy to survive. They might even devise a way to survive proton decay. He believes that, despite the problems, life in an open universe might survive indefinitely.

CHAPTER 16

Epilogue

This concludes our journey to the end of time, and our look at one of the most important problems of astronomy—the dark matter problem. We have seen that according to present ideas most of our universe is made up of a strange, as yet unidentified form of matter. And we have looked at what it might be. In addition, we have seen how this matter determines the fate of the universe. The question that now comes to mind is: How close are we to finally solving the problem? Also, is the picture we have today of the dark matter and the fate of the universe likely to change? It is, of course, always dangerous to try to predict the future. Changes will no doubt come, but it's hard to say what form they will take.

Talking about the future of science always reminds me of one of Sidney Harris's cartoons. In it two scientists are shown looking at some equations on a blackboard. One is saying to the other, "What is most depressing is the realization that everything we believe now will be disproved in a few years."

I don't think everything we believe about dark matter will be disproved, but we may be in for some surprises. As we saw earlier, it is quite possible that there is no dark matter problem. The puzzle may be solved without the introduction of any type of matter. Of course, few scientists believe that this will happen. Still, it's possible.

Despite the problems, considerable progress has been made, and much more is likely to be made in the next few years.

A large number of experiments are presently in progress: searches for exotic particles, proton decay experiments, searches for monopoles and for evidence of massive neutrinos. Also, large, complex computer programs are being written in an attempt to determine what particles give the best match to the known overall structure of the universe. Important results will no doubt come from these experiments. And as large telescopes are put into space and giant accelerators are built, further progress will come. But an all-encompassing breakthrough that solves the problem in all its aspects is not likely to come. Various aspects of the problem will no doubt be solved over the next few years. But it may be several generations before we have anything close to a full solution.

Glossary

Absolute brightness   A measure of the actual brightness of a star, independent of the distance of the star.

Antiparticle   Corresponding to every type of particle there is an antiparticle. When an antiparticle and a particle meet they annihilate one another with the release of energy.

Antiproton   Antiparticle of the proton.

Axion   Particle that is predicted by grand unified theory. Very light.

Baryon   A heavy particle. Made up of three quarks.

Binary quasar   A double quasar. Two quasars that revolve around one another.

Blueshift   A shift of the spectral lines toward the blue end of the spectrum. Indicates an approaching object.

Bottom-up theory   Theory that assumes the galaxies form first, clusters later.

Brown dwarf   An object with slightly less than the amount of mass needed to create a star.

CCD   Charge-coupled device. A device for enhancing images.

Celestial mechanics   The study of the motions of celestial objects.

Closed universe   A universe in which the recession of the galaxies eventually stops. Positively curved.

Cloud chamber   A device that allows us to see the tracks of particles. Ions condense along the tracks making them visible.

Cluster   A group of stars, or galaxies.

Cold dark matter   Dark matter of low velocity (e.g., axions, photinos).

Collider   A particle accelerator designed so that particles moving in opposite directions can collide.

Conservation of energy   A law stating that energy does not change in a physical process.

Constellation   A group of stars that appears to be close to one another in the sky.

Cosmic background radiation   Radiation that was released in the early universe. Now fills the universe uniformly. Has a temperature of 3 K.

Cosmic string   Hypothetical string in the early universe. May be responsible for structure.

Cosmological constant   A constant term added by Einstein to his field equations to make the universe static.

Cosmology   Study of the structure and evolution of the universe.

Critical density   Density at which the universe is flat. Dividing line between open and closed universes.

Dark matter   Matter that appears to be missing. Matter that is not observed.

Decoupling   The release of radiation from matter in the early universe.

Deuterium   A heavy form of hydrogen. Nucleus contains one proton and one neutron.

Differential equation   A type of equation in calculus.

Doppler shift   The apparent change in wavelength of light due to relative motion between source and observer.

Dwarf star   A star somewhat smaller than our sun.

Dyson sphere   Spherical shell surrounding a star. Used for gathering light of star.

Electric field   Lines of force that surround electric charge.

Electromagnetic radiation Radiation (photons) of various wavelengths, ranging from radio waves to gamma rays.

Electron volt The amount of energy acquired as an electron moves through a potential difference of one volt.

Entropy A measure of the disorder of a system.

Ergosphere Region around a black hole between the static limit and the event horizon.

Escape velocity Velocity needed to completely escape a given gravitational field.

Event horizon Surface of a black hole.

Exclusion Principle Principle stating that electrons cannot occupy the same small region of space.

Exotic particles Hypothetical particles such as axion, photino.

Field equations Set of equations describing various types of fields (e.g., gravitational field).

Fluctuation A slight change in density.

Frequency The number of vibrations per second.

Fusion The fusing, or joining together, of two nuclei to form a more complex nucleus.

Fusion reactor A reactor that operates on the principle of fusion.

General relativity A theory of gravity devised by Einstein in 1915.

Globular cluster A group of a few hundred thousand stars (sometimes a few million).

Grand unified theory (GUT) A theory that attempts to unify the electromagnetic, strong, and weak fields.

Gravitational constant (G) The constant that governs the gravitational field of all objects in the universe.

Gravitational lens Deflection of light by a gravitating object.

Gravitational radius Radius of a black hole.

Gravitino Superpartner of the graviton.

Graviton Particle of the gravitational field.

Greenhouse effect The trapping of radiation between the surface of a planet and a cloud layer above it.

Half-life   Pertains to radioactive decay. Time for half of nuclei to decay to lighter nuclei.

Hawking radiation   Radiation emitted close to event horizon of black hole as a result of tidal forces.

Heat death   Gradual approach of universe to 0 K.

Hot dark matter   Dark matter that has high velocity (e.g., massive neutrinos).

HR diagram   A plot of absolute brightness of stars versus surface temperature.

Image tube   An image intensifying device.

Inflation theory   A theory that predicts the universe underwent a sudden inflation during its early expansion.

Interstellar gas   Gas between the stars.

Ionization   Process of removing (or adding) electrons from (to) atom to make it charged.

Jet   Gaseous streamer that emanates from galaxy or other object.

Keplerian motion   Motion similar to that of the planets of the solar system. Inner objects move faster than outer ones.

Kerr black hole   A spinning black hole.

Lensing   Pertains to gravitational lens. The bending of light rays by a gravitational field.

Light-year   The distance light travels in a year.

Local Group   The group of galaxies to which the Milky Way belongs.

Magnetic bottle   A "container" made up of magnetic field lines.

Mass ratio   Ratio of mass of rocket plus fuel on takeoff to mass at end of journey.

Maunder minimum   A period of few sunspots between approximately 1600 and 1700.

Messier number   A number assigned to approximately 100 of the brightest nebulous objects by Charles Messier.

Missing mass   A term usually used in relation to the universe. Mass needed to make it flat.

ML ratio   The ratio of mass to luminosity, or brightness.

MMT   Multiple-mirror telescope. Telescope consisting of several mirrors that work in conjunction with one another.

Monopole   A heavy particle with either a north or south magnetic pole, but not both.

Morphological   A study of structure or form.

Muon   A particle similar to the electron, but heavier.

Nebula   A gaseous region in space.

Negatively curved universe   An open universe. Will expand forever.

Neutrino   A particle that is believed to be massless (but may have a small mass). Experiences only weak interactions and is electrically neutral.

Neutron star   A star made up of neutrons. Usually only a few miles in diameter.

Nova   A star that experiences a sudden outburst of energy that increases its luminosity a thousandfold.

Nuclear reactor   Reactor powered by nuclear reactions.

Nucleosynthesis   The process of generating nuclei in stars (or early universe).

Nucleus   The massive central part of an atom. In relation to a galaxy it is the central portion.

Open universe   A negatively curved universe. Will expand forever.

Pair production   Production of particle and antiparticle.

Perfect absorber   An object that absorbs all radiation.

Period–luminosity relation   The relation between the period of a Cepheid variable and its brightness.

Photino   Superpartner of photon.

Photoelectric cell   A device for measuring the light of a star.

Photon   A "particle" of light or other electromagnetic radiation.

Plasma   A mixture of charged particles.

PMBH   Planetary-mass black hole.

Positively curved universe   Closed universe. Will eventually collapse.

Positron   Antiparticle of electron.

Positronium　　An "atom" consisting of an electron and a positron.

Primordial black hole　　Black hole that formed as a result of an inhomogeneity in the early universe.

Principle of Uncertainty　　Basic principle of quantum theory that tells us we cannot simultaneously measure such things as momentum and position to a high degree of accuracy.

Quantization　　The process of forming a quantum theory.

Quantum cosmology　　A cosmology involving atomic dimensions.

Quantum electrodynamics　　Quantum theory of electrons and photons.

Quantum field theory　　Quantum theory of a field (e.g., electromagnetic field).

Quantum fluctuation　　A fluctuation, or change in density, over atomic dimensions.

Quantum theory　　The branch of physics that deals with the structure and behavior of atoms (and elementary particles) and their interaction with light.

Quark bag　　A "bag" that contains quarks. Baryons are made up of three quarks in a bag.

Quasar　　Energetic object in the outer regions of the universe.

Radiation　　Photons. Electromagnetic energy.

Radio galaxy　　A galaxy that emits radio waves.

Radio telescope　　Telescope designed to pick up radio waves from the sky.

Recessional velocity　　Speed in a direction away from us.

Red dwarf　　A small red star. Considerably smaller than the sun.

Redshift　　A shift of spectral lines in the direction of the red end of the spectrum. Indicates recession.

Relativity theory　　Einstein's theory of motion. Special relativity describes uniform motion; general relativity describes accelerated motion.

Resolution　　Ability to clearly distinguish two closely associated objects.

Rest mass   Mass of a particle when it is at rest. Mass with no contribution due to particle's speed.

Rotation curve   A plot of the rotational speed of stars versus their distance from the center of the galaxy.

Shadow matter   A form of matter predicted by grand unified theory. There are shadow particles corresponding to all types of particles in our universe.

Singularity   A point of infinite density. A point where the laws of physics break down.

Sky glow   Background glow emanating from the sky.

Spectral line   A line that appears when the light from a star or other object is passed through a spectroscope.

Spectrograph   Instrument for photographing spectra.

Spectroscopy   The analysis of light by separating it by wavelength.

Standard candle   A reference light.

Static limit   Region around a black hole. Inside this region objects cannot remain at rest.

Strong nuclear force   Force of the nucleus. Short-ranged.

Supercluster   A cluster of galactic clusters.

Superconductivity   The flow of electricity under conditions of no electrical resistance.

Supernova   Explosion in which most of a star is blown off into space.

Superpartner   Particles in supergravity theory.

Superstring   Tiny string that is related to the makeup of particles and forces in superstring theory.

Tensor   Generalization of a vector (a quantity with direction and magnitude).

Terraforming   Process of changing the climate and atmosphere of a planet so that they resemble those of Earth.

Thermonuclear furnace   Core of a star. Region where nuclear reactions are going on.

Tidal force   A stretching force. Results from the difference in gravitational pull between two points.

Top-down theory   Theory that assumes superclusters were formed first with clusters and galaxies coming later.

Viking experiments   A series of experiments that were performed on Mars during the Viking landings.

Void   Region of universe that contains few if any galaxies.

White nebula   An older term for galaxy of nebula. An object that appeared fuzzy and white in the telescope.

Bibliography

The following is a list of general and technical references for the reader who wishes to know more about the subject. References marked with an asterisk are of a more technical nature.

Chapter 1: Introduction

Harrison, Edward, *Cosmology* (London: Cambridge University Press, 1981).
Kaufmann, William, *Universe* (San Francisco: Freeman, 1985).
Parker, Barry, *Concepts of the Cosmos* (New York: Harcourt Brace Jovanovich, 1984).
Silk, Joseph, *The Big Bang* (San Francisco: Freeman, 1980).

Chapter 2: The Expanding Universe

Arp, Halton, *Quasars, Redshifts and Controversies* (Berkeley: Interstellar Media, 1987).
Bartusiak, Marcia, *Thursday's Universe* (New York: Times Books, 1986).
Berendzen, Richard; Hart, Richard, and Seeley, David, *Man Discovers the Galaxies* (New York: Science History Publications, 1976).
Ferris, Timothy, *The Red Limit* (New York: Morrow, 1977).
Harrison, Edward, *Cosmology* (London: Cambridge University Press, 1981).
Morris, Richard, *The Fate of the Universe* (New York: Playboy Press, 1982).
Parker, Barry, "The Age of the Universe." Astronomy (July, 1981) 67.

Parker, Barry, *Einstein's Dream* (New York: Plenum, 1986).

Shapley, Harlow, *Through Rugged Ways to the Stars* (New York: Scribner's, 1969).

Silk, Joseph, *The Big Bang* (San Francisco: Freeman, 1980).

Weinberg, Steven, *The First Three Minutes* (New York: Basic Books, 1977).

Whitney, Charles, *The Discovery of Our Galaxy* (New York: Knopf, 1971).

Chapter 3: The Missing Mass Mystery Unfolds

Disney, Michael, *The Hidden Universe* (New York: Macmillan, 1984).

Krauss, Lawrence, "Dark Matter in the Universe." Scientific American (December, 1986) 58.

*Neyman, Jerzy; Page, Thornton, and Scott, Elizabeth, "Conference on the Instability of Systems of Galaxies." The Astronomical Journal (November, 1961) 533.

*Oort, Jan, "The force exerted by the stellar system in the direction perpendicular to the galactic plane and some related problems." Bulletin of the Astronomical Institute of the Netherlands (August, 1932) 249.

Parker, Barry, "The Missing Mass Mystery." Astronomy (November, 1984) 9.

Trimble, Virginia, "Existence and Nature of Dark Matter in the Universe." Annual Review of Astronomy and Astrophysics (1987) 425.

Trimble, Virginia, "Our Cosmic Horizon: The Search for Dark Matter." Astronomy (March, 1988) 18.

Tucker, Wallace, and Tucker, Karen, *The Dark Matter* (New York: Morrow, 1988).

*Zwicky, Fritz, "On the Masses of Nebulae and Clusters of Nebulae." The Astronomical Journal (October, 1931) 217.

Zwicky, Fritz, *Morphological Astronomy* (Berlin: Springer-Verlag, 1957).

Chapter 4: Dark Matter in Galaxies?

Disney, Michael, *The Hidden Universe* (New York: Macmillan, 1984).

Faber, S. M., and Gallagher, J. S., "Masses and Mass-to-Light Ratios of Galaxies." Annual Review of Astronomy and Astrophysics (1979) 135.

*Ostriker, Jeremiah, and Peebles, James, "A Numerical Study of the Stability of Flattened Galaxies: Or, Can Cold Galaxies Survive?" The Astrophysical Journal (December, 1973) 467.

*Ostriker, Jeremiah; Peebles, James, and Yahil, Amos, "The Size and Mass of Galaxies and the Mass of the Universe." The Astrophysical Journal (October, 1974) L1.

Rubin, Vera, "Dark Matter in Spiral Galaxies." Scientific American (June, 1983) 96.

Rubin, Vera, "The Rotation of Spiral Galaxies." Science (June 24, 1983) 1339.

Tucker, Wallace, and Tucker, Karen, The Dark Matter (New York: Morrow, 1988).

Chapter 5: Dark Matter in the Universe?

*Applegate, James; Hogan, Craig, and Scherrer, Robert, "Cosmological Baryon Diffusion and Nucleosynthesis." Physical Review D (February, 1987) 1151.

Burns, Jack, "Dark Matter in the Universe." Sky and Telescope (November, 1984) 397.

Disney, Michael, The Hidden Universe (New York: Macmillan, 1984).

Harrison, Edward, Cosmology (London: Cambridge University Press, 1981).

Kaufmann, William, Relativity and Cosmology (New York: Harper & Row, 1973).

Parker, Barry, "The Missing Mass Mystery." Astronomy (November, 1984) 6.

Parker, Barry, Creation (New York: Plenum, 1988).

Chapter 6: Will the Candidates Please Stand

Bartusiak, Marcia, Thursday's Universe (New York: Times Books, 1986).

Disney, Michael, The Hidden Universe (New York: Macmillan, 1984).

Parker, Barry, "The Search for Gravitational Waves." Encyclopaedia Britannica Yearbook: Science and the Future (1981) 182.

Parker, Barry, "The Cosmic X-Ray Background." Astronomy (February, 1986) 90.

Tarter, Jill, "Brown Dwarfs and Black Holes." Astronomy (April, 1978) 18.

Trimble, Virginia, "Existence and Nature of Dark Matter in the Universe." Annual Reviews of Astronomy and Astrophysics (1987) 425.

Tucker, Wallace, "The Missing Mass Mystery." Science Digest (September, 1981) 18.

Tucker, Wallace, and Tucker, Karen, "Stalking the Magnetic Monopole." Mercury (March–April, 1983) 39.

Waldrop, Mitchell, "Massive Neutrinos: Masters of the Universe?" Science (January, 1981) 470.

Chapter 7: Black Holes

Bartusiak, Marcia, Thursday's Universe (New York: Times Books, 1986).

Bekenstein, Jacob, "Black Hole Thermodynamics." Physics Today (January, 1980) 24.

*Freese, Katherine; Price, Richard, and Schramm, David, "Formation of Population III Stars and Galaxies with Primordial-mass Black Holes." The Astrophysical Journal (December, 1983) 405.

Hawking, Stephen, A Brief History of Time (New York: Bantam, 1988).

Hawking, Stephen, and Israel, Werner, 300 Years of Gravitation (London: Cambridge University Press, 1987)

Morris, Richard, The Fate of the Universe (New York: Playboy Press, 1982).

Parker, Barry, "Mini Black Holes." Astronomy (February, 1977) 26.

Parker, Barry, Einstein's Dream (New York: Plenum, 1986).

Parker, Barry, "In and Around Black Holes." Astronomy (October, 1986) 6.

Chapter 8: Neutrinos with Mass

Asimov, Issac, The Neutrino (New York: Doubleday, 1966).

Morris, Richard, The Fate of the Universe (New York: Playboy Press, 1982).

Schechter, Bruce, "A Prodigal Particle." Discover (March, 1981) 20.

Waldrop, Mitchell, "Massive Neutrinos: Masters of the Universe?" Science (January, 1981) 470.

Chapter 9: Magnetic Monopoles and Other Exotic Particles

Carrigan, Richard, and Trower, Peter, "Superheavy Magnetic Monopoles." Scientific American (April, 1982) 106.
Ford, Kenneth, "Magnetic Monopoles." Scientific American (December, 1963) 122.
Schechter, Bruce, "The Hunting of the Monopole." Discover (July, 1982) 68.
Thomsen, Dietrick, "To Catch a Monopole." Science News (December 4, 1982) 362.
Tucker, Wallace, and Tucker, Karen, "Stalking the Magnetic Monopole." Mercury (March–April, 1983) 39.

Chapter 10: Galaxy Formation and Dark Matter

Bartusiak, Marcia, Thursday's Universe (New York: Times Books, 1986).
Finkbeiner, Ann, "Fossils of Something Interesting: The Large-Scale Structure of the Universe." Astronomy (November, 1984) 18.
Finkbeiner, Ann, "Cold Dark Matter and the Origin of Galaxies." Astronomy (April, 1985) 67.
Hut, Piet, and White, Simon, "Can a Neutrino-Dominated Universe be Rejected?" Nature (August, 1984) 637.
Krauss, Lawrence, "Dark Matter in the Universe." Scientific American (December, 1986) 58.
Melott, Adrian, "Cosmology on a Computer." Astronomy (June, 1983) 66.
Melott, Adrian, "Cosmology on a Computer II." Astronomy (July, 1983) 66.
*White, Simon, and Daqing, Liu, "The Origin and Evolution of Structure in a Universe Dominated by Cold Dark Matter." Proceedings of the Yellow Mountain School of Physics and Astrophysics (Singapore: World Scientific, 1988).
*White, Simon; Frenk, Carlos, and Davis, Marc, "Clustering in a Neutrino-Dominated Universe," The Astronomical Journal (November, 1983) L1.

Chapter 11: Gravitational Lenses and Dark Matter

Chaffee, Frederic, "The Discovery of the Gravitational Lens." Scientific American (November, 1980) 70.

Gott, Richard, "Gravitational Lenses." American Scientist (March–April, 1983) 150.

Turner, Edwin, "Gravitational Lenses." Scientific American (July, 1988) 54.

*Tyson, Anthony; Seitzer, P., Weymann, Ray, and Foltz, Craig, "Deep CCD Images of 2345+007: Lensing by Dark Matter." The Astronomical Journal (June, 1986) 1274.

Chapter 12: A Consensus: To Be or Not To Be

*Bahcall, John; Piran, T., and Weinberg, Stephen, *Dark Matter in the Universe* (Singapore: World Scientific, 1987).

Krauss, Lawrence, "Dark Matter in the Universe." Scientific American (December, 1986) 58.

*Milgrom, Mordehai, "A Departure from Newtonian Dynamics at Low Accelerations as an Explanation of the Mass Discrepancy in Galactic Systems." *Dark Matter in the Universe* (Singapore: World Scientific, 1987).

Peebles, James, and Silk, Joseph, "A Cosmic Book." Nature (October 13, 1988) 601.

Chapters 13 and 14: Fate of the Open and Closed Universes

Darling, David, "Deep Space: The Fate of the Universe." Astronomy (January, 1986) 6.

Davies, Paul, *The Runaway Universe* (New York: Harper & Row, 1978).

Dicus, Duane; Letaw, John; Teplitz, Doris, and Teplitz, Vigdor, "The Future of the Universe." Scientific American (March, 1981) 90.

Dyson, Freeman, "Time Without End: Physics and Biology in an Open Universe." Reviews of Modern Physics (July, 1979) 447.

Islam, Jamal, "The Ultimate Fate of the Universe." Sky and Telescope (January, 1979) 13.

Islam, Jamal, *The Ultimate Fate of the Universe* (London: Cambridge University Press, 1980).

Morris, Richard, *The Fate of the Universe* (New York: Playboy Press, 1982).

Page, Don, and McKee, Randall, "The Future of the Universe." Mercury (January–February, 1983) 17.

Parker, Barry, *Einstein's Dream* (New York: Plenum, 1986).

Chapter 15: The Future of Civilization and Life

Abell, George, "The Search for Life Beyond Earth: A Scientific Update." *Extraterrestrial Intelligence: The First Encounter* (Buffalo: Prometheus Books, 1976).

Islam, Jamal, "The Ultimate Fate of the Universe." Sky and Telescope (January, 1979) 13.

Oberg, James, "Terraforming." Astronomy (May, 1978) 6.

Page, Don, and McKee, Randall, "The Future of the Universe." Mercury (January–February, 1983) 17.

Papagiannis, Michael, "The Search for Extraterrestrial Civilizations: A New Approach." Mercury (January-February, 1982) 12.

Papagiannis, Michael, "Bioastronomy: The Search for Extraterrestrial Life." Sky and Telescope (June, 1984) 508.

Ronan, Colin, *Deep Space* (New York: Macmillan, 1982).

Rood, Robert, and Trefil, James, *Are We Alone?* (New York: Scribner's, 1981).

Tipler, Frank, "Extraterrestrial Beings Do Not Exist." Physics Today (April, 1981) 9.

Tipler, Frank, "The Most Advanced Civilization in the Galaxy is Ours," Mercury (January–February, 1982) 5.

Index